SOIL MANAGEMENT STRATEGIES

SOIL MANAGEMENT STRATEGIES

THRIVING TO OPTIMISE FRUIT PRODUCTION

Edited by

ANIL SHARMA
Senior Scientist (Soil Science)
Central Institute of Temperate Horticulture
K.D. Farm, Old Air Field
P.O. Rangreth, Srinagar-190007 **J&K**

NAZEER AHMED
Director
Central Institute of Temperate Horticulture
K.D. Farm, Old Air Field
P.O. Rangreth, Srinagar-190007 **J&K**

SARVENDRA KUMAR
Scientist (Soil Science)
Central Institute of Temperate Horticulture
K.D. Farm, Old Air Field
P.O. Rangreth, Srinagar-190007 **J&K**

JAVID IQBAL MIR
Scientist (Biotechnology)
Central Institute of Temperate Horticulture
K.D. Farm, Old Air Field
P.O. Rangreth, Srinagar-190007 **J&K**

KALYANI PUBLISHERS

LUDHIANA • NEW DELHI • NOIDA (U.P.) • HYDERABAD • CHENNAI
KOLKATA • CUTTACK • GUWAHATI • KOCHI • BENGALURU

KALYANI PUBLISHERS

Head Office

B-1/1292, Rajinder Nagar, **Ludhiana-**141 008 • Ph : 0161-2760031
E-mail : kalyanibooks@yahoo.co.in

Administration Office

4779/23, Ansari Road, Daryaganj, New Delhi-110 002
Ph : 011-23271469, 23274393, 23278688 **E-mail :** kalyani_delhi@yahoo.co.in

Works

B-16, Sector-8, **NOIDA** (U.P.)

Branches

1, Mahalakshmi Street, T. Nagar, **Chennai-**600 017 • Ph : 044-24344684
110/111, Bharatia Towers, Badambadi, **Cuttack-**753 009 (Odisha) • Ph : 0671-2311391
3-5-1108, Narayanaguda, **Hyderabad-**500 029 • Ph : 040-24750368
10/2B, Ramanath Mazumdar Street, **Kolkata-**700 009 • Ph : 033-22416024
Arunalaya, 1st Floor, Saraswati Road, Pan Bazar, **Guwahati-**781 001 • Ph : 0361-2731274
Koratti Parambil House, Convent Road, **Kochi-**682 035 • Ph : 0484-2367189
No. 24 & 25, 1st Floor, Hameed Shah Complex, Cubbonpet Main Road, **Bengaluru-**560 002

KPP K 16476 05

Typesetting at
Kalyani Publishers

ISBN 978-93-272-4795-4

PRINTED AT

B.B. Press, A-37, Sec-67, Noida-201301
and published by Mrs. Usha Raj Kumar for
Kalyani Publishers, New Delhi-110 002

PREFACE

The Book entitled "Soil Management Strategies : Thriving to optimise fruit production" is targeted to help researchers, scholars, students, planners and policy makers to understand the diverse and advance soil management techniques to optimize fruit production. Most importantly, it gives guidance to frame prediction picture before going for fruit crops in some particular area. Experts of international reputation have contributed their experiences as well as their studies in this book which will act as a multifaceted soil management information bank. The book is covering almost all the soil management techniques related to fruit production and soil quality enhancement. The book is an effort to extend the art of soil management in fruit production, not only to scientific community but to the farming community through line departments and extension workers. The book and the topic it is covering will be of great significance to students studying agriculture related subjects like soil science, agronomy, fruit science etc.

The book intend to assemble all the soil management strategies in order to construct a realistic and useful scaffold for appraising the potential effects of management practices on fruit production and soil quality. Principal audiences of the book will be scientists, scholars, students, teachers, professors, scientific officers from private and public sector, progressive farmers, planners and policy makers. This book is covering all the aspects of soil management for optimizing fruit production. Besides the latest information from the scientists working in the field of soil management and fruit production has also been provided. This book will help readers to understand the various strategies that have been introduced in soil science for sustainable fruit production.

Srinagar, 2015

ANIL SHARMA
NAZEER AHMED
SARVENDRA KUMAR
JAVID IQBAL MIR

CONTENTS

LIST OF CONTRIBUTORS

ABHIJIT DEBNATH
SMS (Horticulture), KVK North Tripura, Panisagar

ANIL K. CHOUDHARY
Senior Scientist, Division of Agronomy, Indian Agricultural Research Institute, New Delhi, India

ANIL KUMAR
Asstt. Professor (Soil Science), Lovely Professional University, Jalandhar, Punjab, India.

ANIL SHARMA
Senior Scientist (Soil Science) Central Institute of Temperate Horticulture (ICAR), K.D. Farm, Old Air Field, Rangreth, Srinagar-190007, Jammu and Kashmir

G.P. JUYAL
Head (Hydrology and Engineering), Central Soil and Water Conservation Research and Training Institute, Dehradun

G.R. MARUTHI SANKAR
Principal Scientist (Statistics), All India Coordinated Research Project for Dryland Agriculture, CRIDA (ICAR), Hyderabad

JAVID IQBAL MIR
Scientist (Biotechnology) Central Institute of Temperate Horticulture (ICAR), K.D. Farm, Old Air Field, Rangreth, Srinagar-190007, Jammu and Kashmir

K.L. SHARMA
National Fellow and Principal Scientist (Soil Science), Division of Resource Management, Central Research Institute for Dryland Agriculture, (ICAR), Santhoshnagar Saidabad, Hyderabad-500 059

N.N. REDDY
Principal Scientist (Horticulture), (CRIDA), Hyderabad

NAZEER AHMED
Director, Central Institute of Temperate Horticulture (ICAR), K.D. Farm, Old Air Field, Rangreth, Srinagar-190007, Jammu and Kashmir

O.C. SHARMA
Senior Scientist (Soil Science) Central Institute of Temperate Horticulture (ICAR), K.D. Farm, Old Air Field, Rangreth, Srinagar-190007, Jammu and Kashmir

P.K. MISHRA
Director, Central Soil and Water Conservation Research and Training Institute, Dehradun, 248195

SANAY ARORA
Senior Scientist (Soils), Central Soil Salinity Research Institute, (ICAR)

SARVENDRA KUMAR
Scientist (Soil Science) Central Institute of Temperate Horticulture (ICAR), K.D. Farm, Old Air Field, Rangreth, Srinagar-190007, Jammu and Kashmir

SOVAN DEBNATH
Scientist (Soil Science) Central Institute of Temperate Horticulture (ICAR), K.D. Farm, Old Air Field, Rangreth, Srinagar-190007, Jammu and Kashmir

V.K. SURI
Former Vice-Chancellor, CSA University of Agriculture and Technology, Kanpur, UP, India

1

Soil and Water Conservation Measures in Fruit Plantation

P.K. MISHRA AND G.P. JUYAL

The fruit crops have assumed an important role in the diversification of agriculture so as to optimise the use of varied soils and ecological situations. Fruit cultivation has shown its promise as one of the most remunerative diversification of land use in hill and mountain ecosystem as it provides congenial and varied agro-climatic conditions varying from subtropical to temperate ones. Fruit plantation is also ecologically recommended as an ideal landuse for steep undulating slopes of the hills, otherwise unsuitable for cereal crops.

Rainfed horticulture has assumed greater attention due to better economic returns from relatively poorer lands. Fruit trees provide a better substitute in areas where cropping may not be possible due to limitation of soil, slope, non-availability of irrigation etc.

The degraded lands in rainfed areas generally earmarked for fruit cultivation lack fertile soil and moisture required for establishment and effective growth of plants. Hence, soil and moisture conservation is a precondition for survival of plantations. As majority of rainfed farmers are marginal or small with low risk bearing capacity, the long gestation period of fruit plants is a stumbling block. To overcome this, research efforts have been made to introduce sandwich agriculture or intercropping (millets and legume) into the system (Das *et al.*, 1993).

Alternate land uses are planned for efficient use of resources for sustained long term productivity. The lands with capability classes IV and above having steep slope and poor moisture retention characteristics are recommended to be utilized under some kind of perennial vegetation including fruit crops. Also, the flat areas with lower rainfall (having good moisture holding capacity) can be identified for alternate land use programme. The concept of alternate land use is more appropriate in the arid and semi-arid regions where natural regeneration of vegetation is constrained mainly due to erratic rainfall distribution and poor soil water holding capacity. Also in hilly and mountainous areas, otherwise unsuitable for cereal crops due to topographical and other constraints, fruit cultivation has been recommended for resource conservation and productivity enhancement.

The degraded rainfed lands earmarked for fruit production generally lack sufficient moisture and may suffer from lack of fertile soil or prone to soil erosion. Therefore, taking up appropriate soil and water conservation (SWC) interventions becomes imperative and a precondition before taking up actual plantation work for its successful establishment and future growth. This paper, therefore, discusses some of the prominent SWC interventions needed for fruit production. Full benefits of SWC interventions are realised when they are implemented on watershed basis with active participation of beneficiaries.

SOIL AND WATER CONSERVATION MEASURES

The success of a properly planned alternate land use system, such as fruit cultivation, largely depends on the soil and water resource conservation programme (Lundqvist, 1992 and Hudson, 1987). As the topography of most degraded lands does not permit for adequate water infiltration, it is essential to have some measures to arrest the water and save the thin soil layer from displacement. In the strict sense, such obstructions to the natural flow of runoff water are termed as the SWC interventions. By this the soil loss is reduced. In addition to this, the SWC interventions in the upper region of a topo-sequence help in recharging the underground reservoir (Sharma and Mishra, 1994). Above all, SWC interventions are the first and the foremost in any land use planning from point of view of resource conservation (soil, water and vegetation) and later resource generation. Experience has shown that in extremely degraded lands, natural conservation processes become very slow. In such situations, SWC interventions improve soil moisture status and promote ecological succession; thus providing better vegetative biomass for resource conservation as well as ensuring production on sustained basis.

The SWC measures are broadly classified into two categories (*i*) Vegetative or Biological, and (*ii*) Mechanical or Engineering or Structural. The combination of biological and mechanical measures is called bio-engineering measures. In situations where biological measures only are insufficient to check erosion to a desired level due to high velocity of runoff or discharge, mechanical measures are needed. The mechanical measures not only check soil erosion but also conserve moisture for crop growth. Basically these measures constitute a series of mechanical barriers constructed across the slope to reduce or break the length of slope only and/or both the length and degree of slope to dissipate the energy of flowing water. The important principles to be kept in view while planning structural measures are :

(*i*) Increasing the time of concentration of runoff and thereby allowing more of it to be absorbed and held in the soil.

(*ii*) Intercepting a long slope into several short ones so as to maintain less than a critical velocity for the runoff water.

(*iii*) Protection against damage due to excess runoff.

The selection and the suitability of engineering control measures for different locations depend on soil depth, rainfall, land slope and crops to be raised.

It is unrealistic to have all standard recommendations/specific procedures for SWC interventions for adoption in different soil and climatic environments. The philosophy behind the introduction of any system lies on the consideration of better soil and moisture conservation practices. After carefully analyzing the following relevant factors, and giving due respect to the wisdom of the local people involved in the activities of greening the degraded land, the SWC interventions should get priority to achieve success in the alternate land use programme.

TECHNICAL FACTORS

Rainfall : Amount, intensity, temporal and spatial variability.

Soil : Soil type, soil depth and their spatial variability, fertility status, quality of subsoil etc.

Topography : Landslope, undulations, extent and nature of rock outcrops etc.

Landuse : Natural vegetation and crop cover.

Resource conservation : Needs of food, fodder, fuel, fibre and fruit for the local people.

Social : The local traditional technology must be properly weighed and incorporated with the scientifically designed technology before programme implementation. The technology should be socially acceptable and implemented in active participation of local community.

Economics : Cost effectiveness (Benefit : Cost ratio) of technology and its economic viability, sufficient gestation period should be considered to fully activate the system for optimum output.

IMPLEMENTATION OF SOIL AND WATER CONSERVATION MEASURES

Integrated land use planning including horticulture plantation and implementation of SWC measures can be done in the best way by following a watershed approach.

Watershed, a hydrological unit of an area draining to a common outlet point, is recognised as an ideal unit for planning and development of land, water and vegetation resources, including fruit plantations.

All the SWC interventions for fruit plantations can broadly be grouped under the following categories :

1. Engineering/mechanical measures
2. Vegetative measures
3. Micro-site improvement
4. Water harvesting

ENGINEERING/MECHANICAL MEASURES

The engineering/mechanical measures can be sub-grouped into two categories *viz.* (*i*) hardware and (*ii*) software land treatments (Sharma and Mishra, 1994). In case of hardware treatments (permanent land measures), mostly the measures which modify runoff flow patterns permanently, are considered. Software treatments, mostly semi permanent or temporary in nature, are followed to support the hardware treatments. The hardware treatments are costly and labour intensive activities generally requiring sufficient government funding. The software treatments include site specific measures like microsite improvement to help in better establishment and growth of vegetation. Being cheaper compared to mechanical measure, these can be taken up by the beneficiaries while opting for alternate landuse measures. Different major conservation practices coming under the hardware and software treatments are discussed below.

Hardware Treatments

Contour trenching

Contour trenching is a widely used practice for moisture conservation in plantation and grassland development. It is a practice of excavating trenches along a uniform level across the slope of land. Bunds are formed along the trenches on the downstream side with material taken out of them. The expected service life of a trench is about 3 to 4 years during which time the vegetation is supposed to take up the conservation function.

These are also called infiltration ditches, which are dug across the major land slope to collect surface runoff to saturate the soil environment and induce recharge of local ground water table (Sivanappan and Aiyasamy, 1992). The trenches can be used in areas receiving over 400 mm of annual rainfall and on slopes exceeding 10 per cent with relatively impermeable soils (Weber and Hoskin, 1983).

Contour trenches break the velocity of runoff and store whole or a part of runoff. Trenches should be designed to store 60-70 per cent of runoff from 4 years 6 hrs storm in coarse textured soils (Samra *et al.*, 2004). In fine textured soils about 50 per cent of the runoff water may be stored. The intercepted runoff percolates through the soil slowly and is made available to the plants. Contour trenches may be of two types (Fig. 1.1).

- Continuous, or
- Staggered

Continuous Trenches. The trenches are called continuous when there is no break in length and they can be 10–20 m long across the slope depending upon width of the field.

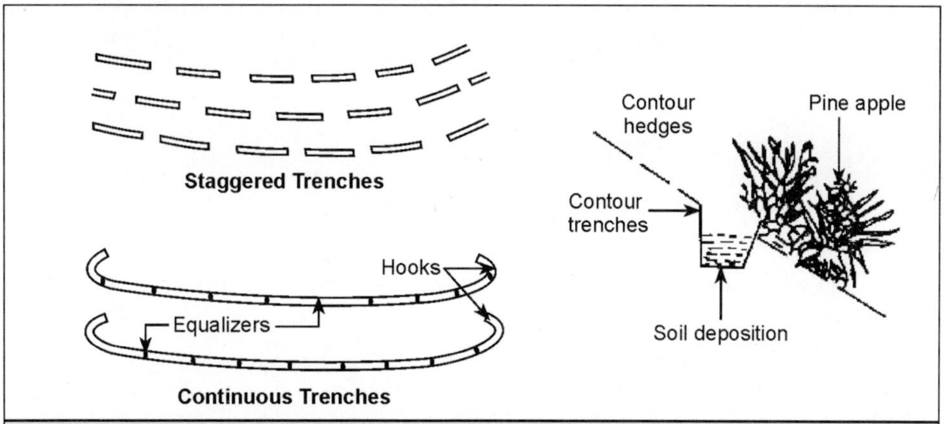

Fig. 1.1. Contour trenches showing staggered and continuous ones and side section

Generally, trenches are dug with a cross-section varying from 30 cm × 30 cm to 45 cm × 45 cm (Fig. 1.2). This type of trenches are generally constructed on mildly sloping uniform slopes.

Fig. 1.2. Continuous contour trenching

Staggered Trenches. The staggered contour trenches are laid scattered with interspace between them (Fig. 1.3). In staggered trenching, the trenches are located directly below one another in alternate rows and in a staggered fashion. Staggered trenches may be made to a length of 2-3 m and spacing between the rows may vary from 3-5 m. This type of trenches are constructed on relatively steeper slopes with irregular topography.

Fig. 3. Staggered contour trenching

Bunds/Dikes/Berms

Bunds are constructed as a series of low height ridges/dikes on contours or on a slight gradient. The purpose is to intercept runoff, reduce soil erosion and temporarily store water to augment infiltration and recharge. Bunds can be used on slopes between 2 and 20 per cent on soils that generate heavy runoff. Small and fragmented land holding is sometimes a bottleneck for construction of bunds and also particularly in areas having thick natural vegetation or rock outcrops.

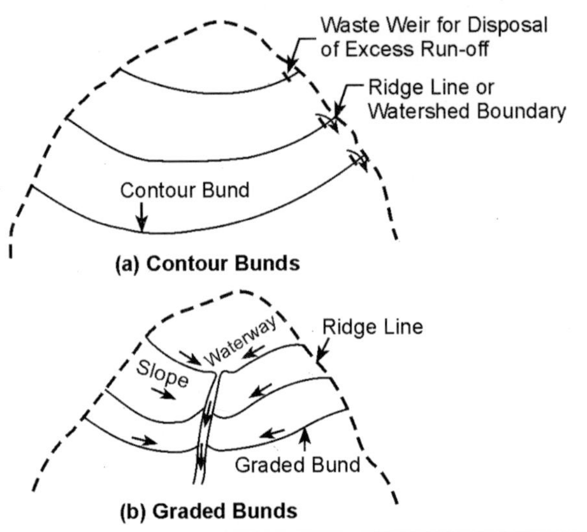

Fig. 1.4. Berms/Bunds

Design of the bunds depends on rainfall, soil and topography. The bunds with lateral gradient (graded bund) discharge runoff to the water way for safe disposal to downstream water bodies. Bunds on contours (contour bunds) should have waste weirs to drain out excess flow from the field and protect the bunds from wash out. Bunds must be maintained before and during the rainy season in the initial years. Line diagrams of contour and graded bunds are shown in (Fig. 1.4).

Orchard Terraces

Orchard terraces, primarily used for establishment of tree crops, are discontinuous type of narrow inward sloped bench terraces which are applicable on slopes up to 30° (58 per cent) (Fig. 1.5). The space between two terraces is determined by the planting distance of the tree crops or for the convenience of the crop management. As they are generally used on steep slopes, the spaces between them should normally be protected by a permanent vegetative cover such as grass, legumes, or other cover crops. The main tree crops are planted in individual basins (Sheng, 1989).

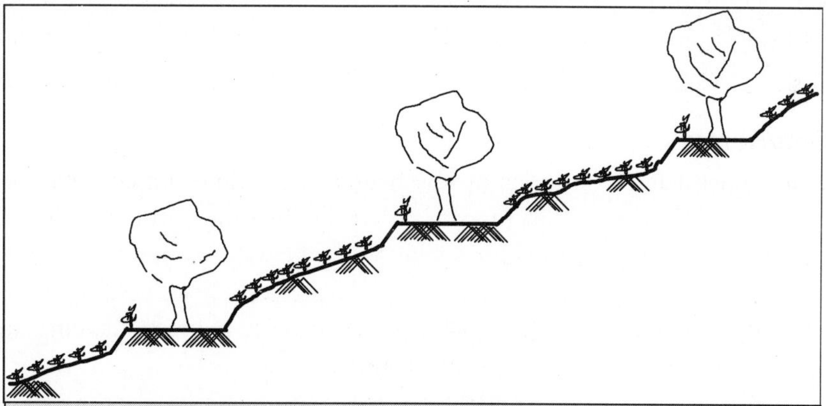

Fig. 1.5. Orchard terraces

Half Moon Terraces

Individual basins in semi-circular shape, also called 'half-moon' terraces, are used for planting individual plants (Fig. 1.6). The diameter of the basins is adjusted according to crop. A shoulder bund at the lower periphery should be provided (Sheng, 1989).

The basins retain soil moisture and prevent fertilizer being washed away. They are particularly useful for establishing semi-permanent or permanent tree crops on steep slopes for erosion control.

Suitability

- They can be constructed on relatively uneven or dissected terrain and on shallow soils.
- Half -moon or Semi-circular terraces with crescent bunds with 0.75 to 1.25 m diameter are found to be effective for soil and moisture conservation especially in low to

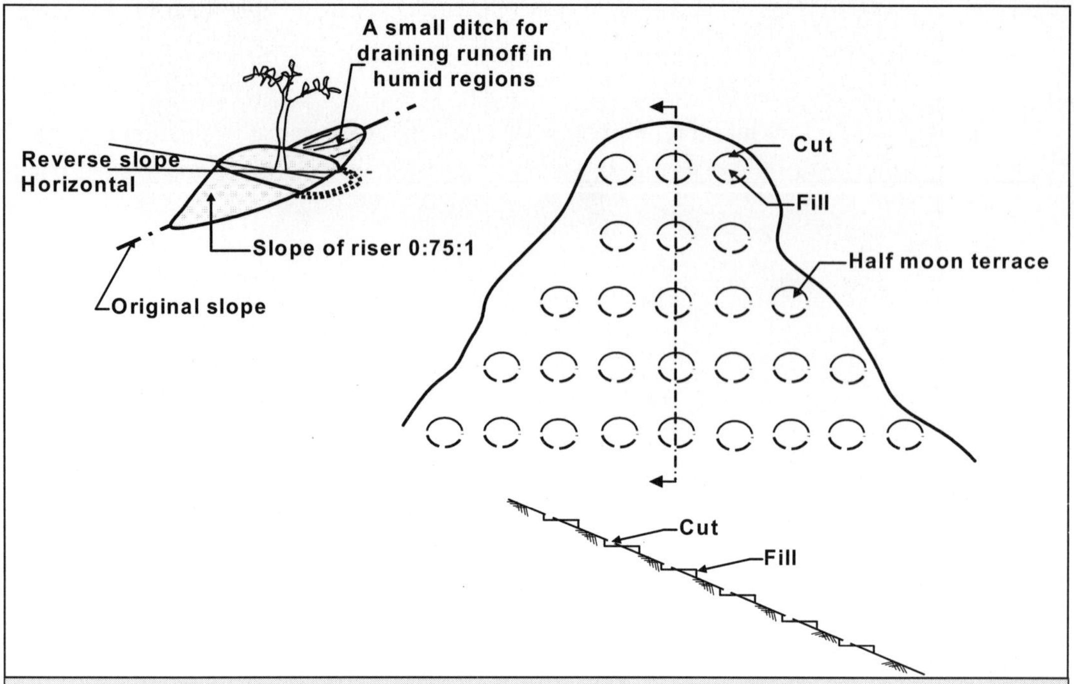

Fig. 1.6. Half moon terraces on hill slopes

moderate rainfall and hilly areas for planting fruit trees and afforestation in moderately deep soils.

Stone Walls

Stone walls can be constructed across the hill slopes at pre-determined spacing for developing land for cultivation, if stones are readily available in the locality (Fig. 1.7 and

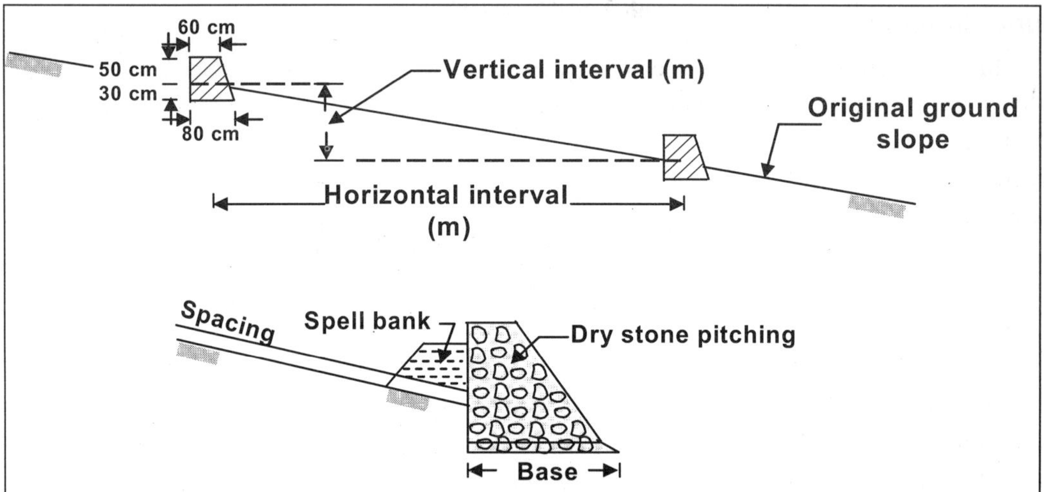

Fig. 1.7. Stone wall

Fig. 1.8. Stone walls for sloping area land treatment (Kerala)

Fig. 1.8). Stone walls can be constructed on contour or at minor gradient depending upon the need (Sharda *et al.*, 2012).

Suitability

- Stone walls are suitable at places where stones are readily available.
- The stone walls help in retaining soil and subsequent formation of a bench terrace as well as drainage of the excess water.
- Contour stone walls are widely adopted in the hilly areas across the country.

The terrace is designed considering the rainfall intensity, landslope and soil infiltration characteristics (Singh *et al.*, 1990). Terracing requires proper design consideration and the construction has to be undertaken by trained professionals.

Gully Stabilization Structures

In upstream area of the watershed, where channel erosion starts by forming gullies, gully stabilization structures should be erected. Priority should be given to active gullies which if not checked may extend into better lands. Properly designed gully check/drop structures when erected in series can stabilize the gully and restore the upstream land by taking up permanent vegetation. In most cases the checks can be built from locally available materials. Towards the downstream of the gully, massive interventions by gabion/masonry structures are provided to dissipate the kinetic energy of running water. Protection of gully banks can also be done by introducing trees, grasses and bushes with strong root systems. Gabions can also be used for gully stabilization. The impounding of water in gullies facilitates ground water recharge. It is desirable to have provision of spillway for disposing of excess water safely. Some commonly used gully stabilization structures are shown in (Figs. 1.9 and 1.10) (Singh *et al.*, 1990).

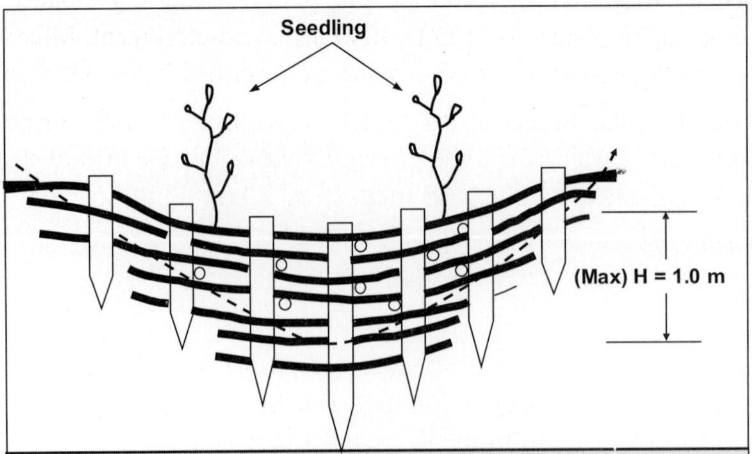

Fig. 1.9. A brush wood check dam

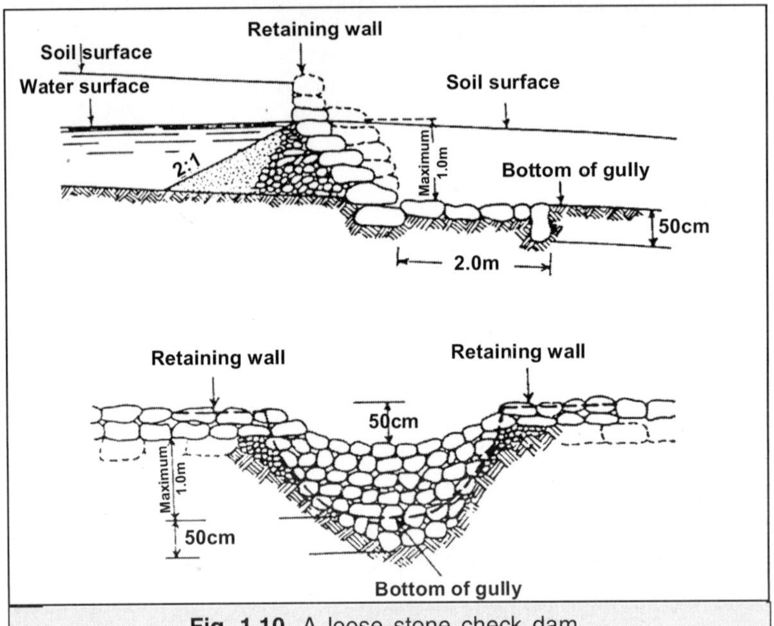

Fig. 1.10. A loose stone check dam

Software Treatments

Micro-catchments. In the plantation works, runoff water harvesting for *in situ* moisture conservation can be exploited for better production. The runoff harvesting should be done in such a way that each tree has its own micro-catchment area. The layout of micro-catchment in relation to the location of tree depends upon the topography of the area. The factors which affect the quantity of runoff are the intensity and duration of rainfall, size of catchment slope of the catchment and its surface characteristics. The optimum size of the micro-catchment should be decided as per the runoff water availability and plant water requirements. Substantial part of post-monsoon rains can also be harvested effectively through land shaping which can avoid supplementary irrigation for a considerable period during dry months. Depending on rainfall, fruit trees can be planted on 125 to 1000 m^2 micro-catchment. Micro-catchment can be given various configurations by land shaping as described below (Sharda *et al.*, 2012).

Ring basins. Circular basins of 0.5 to 1.0 m radius are usually made 7-10 m apart, depending on the type of fruit trees to be grown. Usually, 3 to 5% inward slope is given and a mound of earth is made all around the trunk (Fig. 1.11a).

Square basins. Normally, square basins of 7 × 7 m size are made and the fruit tree is grown in the centre. About 5% slope from all directions is given to induce runoff (Fig. 1.11b).

Rectangular basins. In a rectangular layout of the orchard, basins of either 5 × 7 m or 7 × 10 m size are made and plants are established in the centre of the rectangular basin. About 5% inward slope is given to induce runoff towards trunk (Fig. 1.11c).

Modified micro-catchment. In case of much larger size of micro-catchment (200-1000 m^2) in low rainfall areas, fruit tree is grown in a corner with 5% slope from three directions (Fig. 1.11d).

Cup and saucer. Cup and saucer shaped pits with single or double basin are suitable for fruit crops in dry zone receiving rainfall between 200-500 mm (Fig. 1.11e).

V-shaped catchment. V-shaped micro-catchment with 5% inward slope from either side not only helps in collecting maximum runoff from scanty winter spring rains but also helps in maximum retention of soil moisture in the rooting zone of the plants for a prolonged period. In case of excess rains, V-notch also serves as a drain and avoids water-logging. The size of the catchment varies with spacing of a particular fruit type and rainfall of the area (Fig. 1.11f).

Roaded catchment. It is somewhat like V-shaped catchment but instead of slope from both sides, only 5% slope from one side is provided which induces runoff in tree rows (Fig. 1.11g). Mulching with any biomass in tree rows can also be practised to conserve moisture during dry months.

Contour and staggered trenches. Contour and staggered trenches harvest runoff in the trenches and conserve moisture *in situ* for growing fruit trees on sloping lands in high rainfall areas (Figs. 1.11h and i).

Half-moon terraces. On steep sloping fields under this system, the trees should be planted in half-moon terraced basins having inward slope and bunding over terrace riser protected with bamboo splits or stones. Plantation of fruit trees in half-moon terraces is highly economical and effective in conservation of *in situ* moisture for a considerable period (Fig. 1.11j).

Water Spreading

Sometimes a part of the water flowing in a stream or gully or waterway is diverted to gently sloping lands for ensuring quicker growth of vegetation. The water thus diverted is not

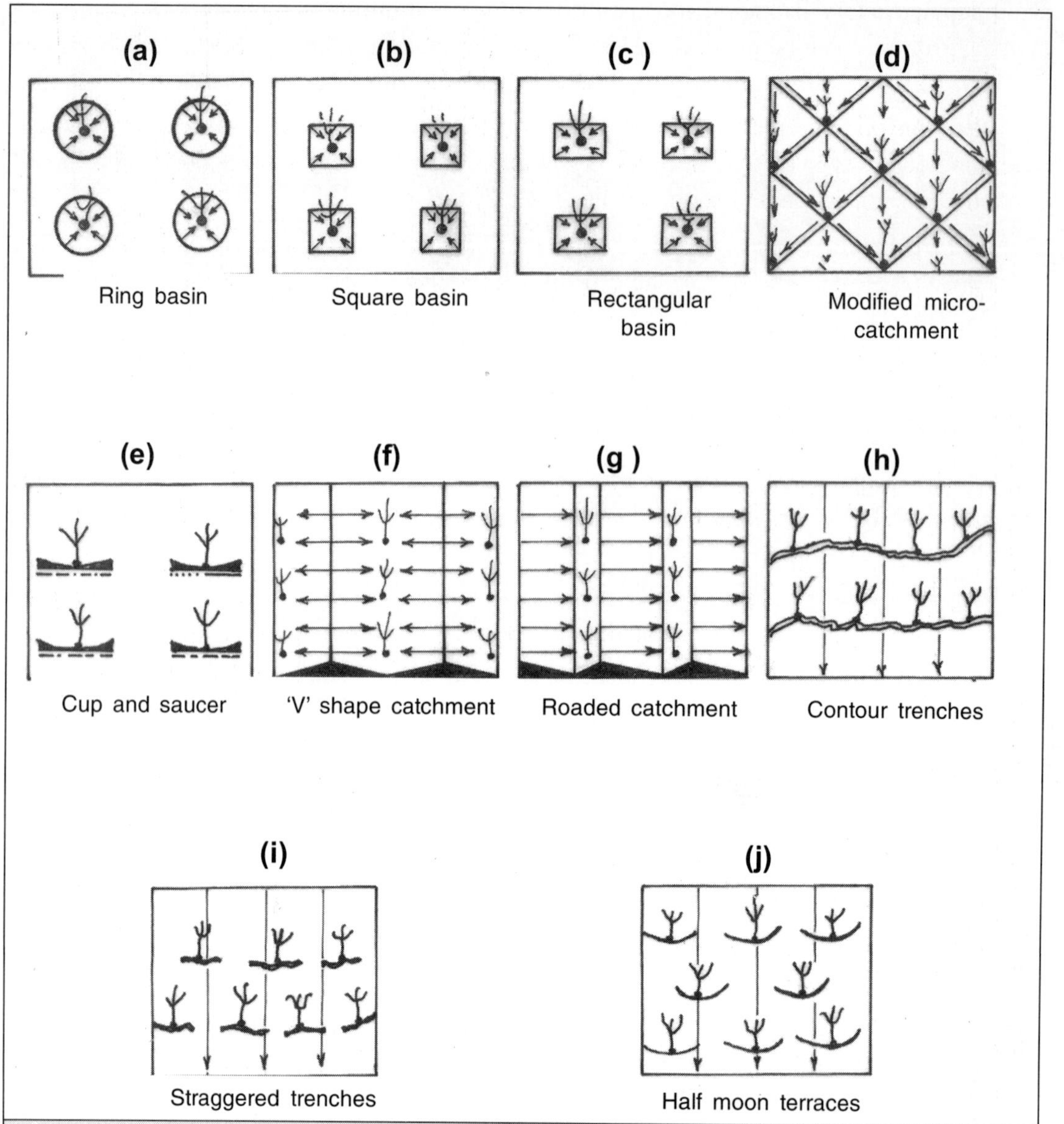

Fig. 1.11. Some land configurations for *in-situ* moisture conservation

allowed to flow through the least path, likely to erode the soil. Hence, interventions in form of berms or trenches are particularly formed to increase the length of flow path of the running water and increase the opportunity time for infiltration.

Soil Working and Pitting Techniques

The soil working techniques depend upon the climate and edaphic factors (Table 1.1). The soil depth and proportion of gravel or boulders at the site, decide the depth and size of pits/trenches. The common soil working techniques are as under (Khanna, 1977).

TABLE 1.1

Type of Pits/Trenches under broad rainfall groups

Type	Rainfall distributor	Rainfall amount		Rainy days		Suitable conditions
		Summer (cm)	Winter (cm)	Summer (No.)	Winter (No.)	
Pits : Ordinary	SH and WL	*25-75*	*Up to 38*	*15-50*	*Up to 15*	Clayey, saline and sandy soils
		SL and WH	Up to 25	25-75	Up to 15	15-30
		SL and WL	Up to 25	Up to 38	Up to 5	Up to 15
		SH and WH	25-75	25-75	15-50	10-30
Saucer			**As above**			Loamy soils, Sloppy land
Ring			**As above**			Dry areas
Ridge-ditch : Vertical Sides						
Small	***SH and WL***	***50-75***	***Up to 38***	**30-40**	**Upto 15**	**Loamy soils. Most parts of dry zone**
	SH and WH	25-75	25-75	**15-50**	10-30	Deep soil
Large	SH and WL	38-50	Up to 38	**30-40**	Upto 15	III- distributed rainfall
	SH and WH	25-50	25-50	**15-50**	15-30	
Shallow	SH and WL		As above			Top soil washed away
	SH and WH					Gravels or murrum exposed
Sides sloped	***SH and WL***	***38-75***	***Up to 38***	**20-50**	**Upto 28**	**Shallow surface soil sub-soil gritty, but fine caly with crusting. More rainy days.**
Slanting sides Side half slanting	***SH and WL***	***25-38***	***Up to 38***	***15-30***	***Up to 15***	**Soils friable**
		Up to 25	Up to 38	Up to 5	Up to 15	Rainfall low
		Up to 25	25-75	Up to 15	15-30	III-distribution
Sides double slanting	***SH and WH***	***50-75***	***Up to 38***	***30-50***	***Up to 15***	**Loamy clay soils**
	SH and WH	25-75	30-50	30-50	10-20	Good well distributed rainfall
Shallow trench	***SH and WH***	***38-75***	***Up to 38***	***20-50***	***Up to 15***	**Alkaline/saline defoculated, clay soils, scanty rainfall**
	SH and WH	25-75	25-75	15-50	10-30	
Deep trench	***SH and WL***	***25-38***	***Up to 38***	***15-30***	***Up to 15***	**High salts, scanty rainfall**
	SL and WL	Up to 25	Up to 38	Up to 5		

SH = Summer high; SL = Summer low; WH = Winter high; WL = Winter low

A. Pits

Ordinary Pit. An ordinary pit is recommended for clayey and saline/alkaline soils in all rainfall classes. It is a pit dug up 30 cm deep with 30 cm base and filled back with earth. It has a high crest and a deep part of the water storage ring, away from the highest point of the mound. This helps in minimising water logging and concentration of salts near the seedlings (Fig. 1.12).

Saucer Pit. This type of pit is suitable for all loamy kinds of soils in all rainfall classes of the dry zone. The crescent trench is necessary only on sloping lands (Fig. 1.12). The pit is made in the centre of 1m radius prepared by giving the soil a slope from periphery towards the centre.

Ring Pit. This is suited for sandy soils in dry areas (Fig. 1.12). The crest is not required to be very high because deep planting will be necessary to place the roots about 30 cm below the surface where a zone of moisture accumulation usually occurs in case of very sandy soils. The ring trench will allow water to infiltrate near the planting site and thus improve its moisture status.

The above-mentioned soil working techniques are adopted for flat lands whereas on sloping lands, ridge ditch is suitable which is partly filled in trench for sowing and planting.

B. Ridge Ditches

Vertical Sides

Small Ridge-Ditch. This is suited for moist part of the dry zone with large number of rainy days and for loamy soils (Fig. 1.12).

Large Ridge-Ditch. This is suitable for low rainfall and deep soil conditions. This traps a large amount of moisture in the worked up soil if the soil is sandy and the rainfall is rather low but the number of rainy days is not small. It is also suitable for low and ill-distributed rainfall for non-sandy soils (Fig. 1.12).

Shallow Ridge-Ditch. This is most suitable for slopes from where top soil is washed away and the surface consists of loose gravel or murram of shallow depth where the soil is extremely granular and has low moisture retentive capacity (Fig. 1.12). Since half rock is encountered near the surface, a large volume of worked-up soil is easily obtained by doing deeper excavation. As the excavated material is stony and gritty, it is advisable to do some sorting of this, and to put the large pieces at the bottom of the trench and at the toe of the ridge, on the lower side of the slope.

Sloped Ridge-Ditch. This method is quite suitable if the surface soil is shallow and the sub-soil is also coarse and gritty but fine clay with crusting nature is present in appreciable amount (Fig. 1.12). This stores adequate moisture in the plug of the worked-up soil and does not create water logging near the ridge.

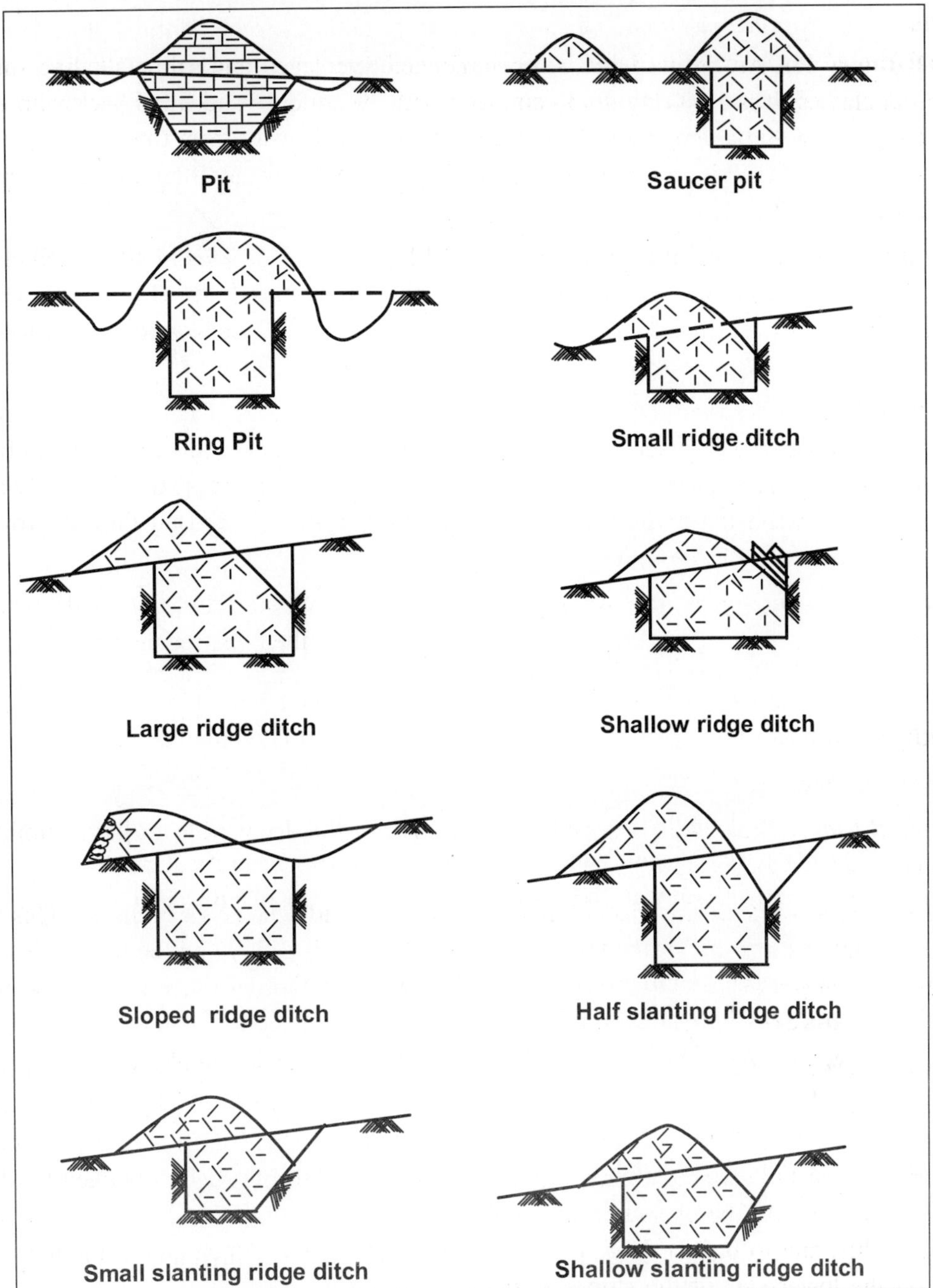

Fig. 1.12. Soil working techniques (diagrammatic)

Slanting Sides. For clayey profiles where water remains standing for a relatively longer period and there is greater seepage of moisture in a vertical direction and roots lie in dry zone, the following techniques may be suitable :

Half Slanting Ridge Ditch. This method may be adopted if the rainfall is low and rather ill-distributed but the soil is friable and does not require much working. This pattern helps in increasing the water collecting space (Fig. 1.12).

Small Slanting Ridge Ditch. This type is suitable for high and well distributed rainfall areas (Fig. 1.12). It prevents water logging.

Shallow Slanting Ridge Ditch. This is suitable for shallow, rather clayey soils with good rainfall. However, the deep slanting ridge-ditch is suitable for gravely, clayey, medium to deep soils with low ill-distributed rainfall (Fig. 1.12).

Double Slanting Ridge-Ditch. Suitable for loamy and clayey soils with good well-distributed rainfall.

Shelved Trench. For peninsular region where clay and murrum form impermeable crust, which on wetting and drying bind granular murrum in hard clod, shelved trenches are recommended. In shelved type of trench, planting is done at ground level or little above it. Shelved trenched may be shallow or deep.

Shallow Shelved Trench. In this technique water will be retained mainly in deeper part of the trench.

Deep Filled Shelved Trench. This is suitable for area with high rainfall which is received in fewer rainy days as heavy or torrential showers. It is recommended for sandy and gravely soils and also for saline and alkaline soils.

Double Trench. This method is suited for stony and detrital slope with small amount of soil in the interspaces of small and large pieces of stones where water holding capacity is very low. It impounds water in the filled portion. Sowing or planting is done at the top of the ridge.

Trench Ridge. It is suitable for salt-affected soils and provides essential mechanism for efficient leaching of salts from the seed bed and impounding of water not in direct contact with the worked up soil and thus minimises damage from water-logging. The shallow trench ridge is suitable for alkaline, saline and deflocculated clay soils in areas receiving rainfall over a large number of rainy days. A deep trench ridge is suited to soil with a high salt content in scanty rainfall received in a few violent storms. It is spacious enough to impound water and has a high ridge to store moisture and allow leaching.

Drilling. Auger holes are made either manually or power driven which economises the cost of soil working. The soil from the holes is excavated through long-armed spatula. Planting in auger hole is, however, suited for the tree species, which have more vertical growth of the tap roots. Soil working should be completed well in advance to allow the excavated soil to weather and the refilled pits get a chance to settle with two or three showers.

Vegetative Measures

Contour planting

The planting should be done along the contour lines to reduce the runoff and soil loss from the sloping land. This technique can be used on most slope situations. This is more important, where annual crops are taken along with the tree components. The planting of trees on contours also serves as key line for taking up regular agricultural operations on lines parallel to the key line.

Vegetative barriers

Characteristics of vegetative barriers. Vegetative barriers, also known as live bunds, are closely-spaced plantations usually of a few rows of grasses or shrubs grown along contours for erosion control in agricultural lands (Figs. 1.13a and Fig. 1.13b). These vegetative barriers not only help in resource conservation but also provide much needed biomass to meet the needs of rural communities.

Design procedure. The design of a vegetative barrier would include selection of suitable species, number of rows *i.e.* spacing and vertical interval. Choice of species is governed by the following characteristics of vegetative barriers (Chandra Prakash *et al.*, 1999).

- Capacity to form erect, stiff, uniform, dense and permanent hedge so as to offer high resistance to overland flow.
- Perenniality so as to serve as a dense hedge for a long time.
- It should not spread as a weed in the crop land.

Fig. 1.13a. A vegetative barrier on agricultural land

Fig. 1.13b. A vegetative barrier in the maize field

- Ability to repel rodents, snakes etc.
- Should have a deep penetrating root system, capable of binding soil effectively.
- It should have the capacity to sprout new tillers quickly after the first rains.
- Should not compete with the crop, plants and there should be minimum loss of crop yield.
- Should be farmer friendly *i.e.* cheap, easy to establish as a hedge and also easily removable if the farmer no longer wants it.
- The biomass/product should have economic value to the farmers.

Evaluation of grass barriers

Different vegetative barriers have been identified for various agro-ecological regions. Some suitable species for grass barriers under different land situations are given in Table 1.2. Vetiver (*Vetivaria zizanioides*) or *khas*, initially promoted as a promising grass species for vegetative barrier by the World Bank, was not found universally effective and acceptable by the farmers, particularly in the arid and semi-arid rainfed areas. The vetiver was inferior to the species of regional importance such as *Eulaliopsis binata* (*Bhabar*) and *Saccharum spontaneum (Kans)* in the Shiwaliks, *Cenchrus ciliaris* in the dry Vertisols, *Pennisetum hohenackeri* in the dry Aflisols, and *Panicum maximum* in the sub-humid lower western Himalayas. In the NEH region, the grasses such as *Cenchrus ciliaris, Panicum antidotale* and *Pennisetum polystachyon* were found effective in reducing soil loss from 40.9 t/ha to < 0.5 t/ha under traditional shifting cultivation.

TABLE 1.2
Choice of vegetative barriers under different land situations

Situations	Species
Marginal soils/desert land/ sand dunes/ravines	*Cenchrus ciliaris* *Cenchrus setigerus* *Chrysopogon fulvus* *Lasiurus sindicus* *Atylosia scarabaeoides* *Lablab purpureus* *Clitoria ternatea* *Stylosanthes humilis* *Stylosanthes hamata* *Macroptilium atropurpureum*
Cultivable wastelands	*Cenchrus ciliaris* *Cenchrus setigerus* *Sehima nervosum* *Dichanthium annulatum* *Chrysopogon fulvus* *Stylosanthes* sp. *Macroptilium atropurpureum* *Clitoria ternatea* *Centrosema pubescens* *Glycine wightii*
Saline and alkaline lands	*Chloris gayana* *Chloris barbata* *Diplachne fusca* *Lasiurus sindicus* *Setaria sphacelata* *Macroptilium atropurpureum* *Eurochloa mosambiensis*
Acidic soils	*Pennisetum polystachyon* *Paspalum notatum* *Centrosema pubescens* *Demodium intortum* *Stylosanthes guyanensis*
Lands affected by shifting cultivation	*Cenchrus ciliaris* *Cenchrus setigerus* *Chrysopogon fulvus* *Clitoria ternatea* *Lablab purpureus* *Stylosanthes scabra*
Marshy lands	*Brachiaria mutica* *B. brizantha*
Irrigated lands	*Panicum maximum* *Pennisetum purpureum* *P. americanum*

In sub-humid conditions of Doon Valley, Guinea grass was found to be most effective in reducing runoff to 32.5% and soil loss to 6.2 t/ha followed by Vetiver (33.9% and 6.6 t/ha) and Bhabar (37.1% and 8.2 t/ha) grasses, respectively compared to 52% and 45 t/ha under cultivated fallow and 48% and 18 t/ha under contour cultivation of maize. The dry matter production was also maximum under Guinea grass (15.4 q/ha) followed by Vetiver (10.9 q/ha) and Bhabar (5.4 q/ha).

In general, vegetative barriers can reduce runoff and soil loss by 18-21% and 23-68%, respectively on slopes varying from 2-8%. Maize and wheat yields increased by about 23-40% and 10-20%, respectively through conserved moisture (Ghosh, 2009). From the system, nearly 6-17 q ha^{-1} dry grass yield is obtained as fodder from the barrier, in addition to maize crop (grain + stalk).

Mixed vegetative barriers of pineapple and grass have been found effective for Western Ghats region, covering the states of Kerala, Tamil Nadu, Karnataka, Maharashtra and Goa where pineapple is commercially cultivated under plantation crops (Madhu and Khola, 2009). Guatemala (*Tripsacumlaxum*) or Hybrid Napier grasses are recommended alongwith two rows of pineapple at 1.0-1.5 m vertical interval.

Vegetative barriers need an appropriate time for establishment and consequent growth, due to which, it takes 2-3 years for these barriers to become effective in contrast to mechanical measures (bunds). On the other hand, the cost of establishing vegetative barriers is much less than for constructing a mechanical barrier. In certain areas, where the slopes are higher, establishment and effectiveness of barriers can be improved by raising a miniature bund on the downstream side. Vegetative barrier technology is highly beneficial for marginal and small farmers since it is easier to establish, easy to maintain, and provides direct benefits like fodder and thatching material of economic value.

Microsite Improvement

The microsite improvement is a form of hydrological intervention for improving point water storage particularly in degraded soil with limited water holding capacity. In degraded soil, perennial vegetation in terms of tree components can be raised successfully through restructuring of the soil at the point where the tree is planted. The restructuring of soil through soil ameliorants can be formed as micro-site improvement. This is nothing but the replacement of native soil with a synthesized mixture of high water holding material. The microsite improvement ensures betters moisture regime around the plant ecosystem by holding more moisture and the same time improving aeration and drainage characteristics. The soil for this purpose can be amended in many ways through ameliorants (organic/inorganics) in different proportions. Also, some sort of mulching either vegetative mulch (straw, grass etc.) or stone mulch should be spread in the tree basin to reduce evaporation losses. The improved microsites act as catch pits for arresting runoff as it increases the amount of rain water storage in soil in the root zone and also allows better recharge of ground water when implemented on a large scale. By this technique a satisfactory survival rate of seedlings can be achieved and vigorous growth of plant is ensured. The amendment of sand/coarse soil with organic and inorganic materials combined with farm yard manure (FYM) offer tremendous scope in improving the soil moisture retention (water holding capacity) characteristics of the microsite and vertical drainage. Some water absorbing polymers (hydrophilic compounds) are now a days available for use in the microsite like : neopeat, soil rite, water works etc. It is always desirable to have bigger size pits for this purpose, provided hard rocks are not encountered at shallow depths. However, cost becomes a limiting factor at times, but it could be reduced by using mechanical pitting augers. Similar, experiences have been recorded with bentonite by Muller and Engelthaler (1985) in Argentina. Other clay minerals available for the purpose are perlite and vermiculite; these materials are being used in nurseries only. In most situations, tank silt is a locally available cheap and practical material. Use of tank silt is nothing but a practice of "soil harvesting and recycling" for benefit of both plant as well as reservoir. In a laboratory study, a mixture consisting of tank silt (33%), alfisol (42%), poultry manure (15%), and paddy husk (10%) improved the soil moisture content by 59% on first day and 237% on the 30th day over the soil (alfisol) alone (Katyal and Osman, 1994). This indicates a prolonged availability of soil moisture by mixing indigenously available cheap materials with native soil. In a shallow alfisol, the addition of bentonite in the pit mixture may be preferred in low dose (Upto 10%) for hardy plants (Mishra and Osman, 1995). However, all the conventional and non-conventional amendments should be tested in the desired environment before final adoption.

Water Harvesting

This is one of the most important component in utilizing the stored water in a part of the catchment area as well as downstream land (Mishra and Sharma, 1992). This technique has wide scope in the areas having moderate to high runoff producing potential. In the initial stages of greening the degraded lands as a part of alternate land use, the availability of stored water is very important from the point of view of off season/supplemental or life saving irrigation. Particularly for raising tree components, arrangements for irrigation should be thought of in the dry periods, at least in the first two years. Hence, provisions must be made to store some water within the easy reach of the farmers, for conveniently using the harvested water, when required. Major storage structures need to be located in community areas or Government lands.

Assessment of runoff and storage capacity and information on soil, land, geology are pre-requisites in designing such structures. As far as possible, local materials should be used for making hydraulic structures for the purpose of impounding and storing water. Depending on the value of harvested water in water scarcity areas, lining of storage structures may also be thought of. The drainage lines (water courses) should be properly aligned and maintained in order to feed the storage structures. Where the runoff amount is more and it is required to store water at number of points; a net work of small water pools may be created along the waterways with offset feeder channels. From the hillock areas water can also be collected by the drain running at the base of the hillock and it is diverted to water harvesting bodies through gullies or water ways.

Polylined Ponds/Tanks

The harvested water from springs/streams is stored in ponds/tanks before application to the crops. To prevent seepage losses, lining of the tanks/ponds is to be resorted to. Conventionally cement lining was being done but since it is very costly an alternate low cost lining by low density polyethylene (LDPE film) (800 to 1000 gauge thickness) is preferred (Juyal and Gupta, 1985). Geomembrane and Silpaulin sheet (200 GSM) have also been used for lining of the small ponds (Sharda and Juyal, 2013). Polylined ponds cost about half of the cement lined ones even after taking into account their respective life periods. Polylined tanks/ponds (20-50 cum storage capacity) are particularly suitable for hilly areas of Himalayan region where cement lined tanks are prone to failure by earthquake and unequal settlement. A series of polylined ponds may be constructed along a stream in the hill areas. The low cost water harvesting poly-ponds can be integrated with gravity fed micro-irrigation system (drip) in the hilly areas for irrigation of high value horticultural crops. The naturally available gravity head in the terraced lands in hills provide an ample opportunity in this regard without resorting to use of external pumping device (Patra *et al.*, 2013).

Drip Irrigation

The drip irrigation can be defined as slow application of water on, above, or beneath the soil surface near or into the plant root zone at the rate equivalent to consumptive use by surface or sub-surface emitters. It is one of the advanced micro irrigation methods becoming popular in our country. However, the concept of micro irrigation is not new in India as traditionally pitchers, porous cups and perforated bamboos were used for irrigation in various parts of the country. Drip system basically consists of main line, sub-mains, laterals, valves, drippers or emitters, riser valves, vacuum breakers, pressure gauges, water meters, filters, fertilizer tanks or venturi pumps, flush valves and pressure regulators. From the main, feeder lines run across the field and laterals are inserted on the same. Low-density polyethylene pipes are laid along the plant or tree rows with outlets or drippers inserted at appropriate intervals. These are designed to supply water at the desired rates (1 to 10 litres/hr) directly to the soil. Low pressures ranging from 0.35 to 2 kg/cm^2 are sufficient to operate the system. Usually, 3.75 to 7.00 cm diameter pipes are used for mains and 2.5 to 3.75 cm for sub-mains. For laterals, pipes of various diameters ranging from 6 to 18 mm with a pressure rating of 4 atmosphere are generally used.

The performance of these systems on horticultural and other row crops has been very encouraging. It has been observed that there is about 50 to 70% water saving and 25 to 30% saving of fertilizers and pesticides when they are applied through the system. For irrigation purposes with advanced technologies, the system can be controlled centrally; hence, also reduces the labour requirements. For efficient water management, demonstration plots for drip irrigation were laid out in coconut, banana and sugarcane in Salaiyur watershed of Coimbatore district (Tamil Nadu) (Fig. 1.14). Drip irrigation under banana is presented in Table 3. Earlier, the farmer used to cultivate only 0.2 ha of banana by surface method of irrigation but now with drip irrigation, 0.4 ha of banana is being cultivated. Water saving of

Fig. 1.14. Drip irrigation in banana field

29 per cent was observed due to drip apart from other benefits such as obtaining produce early and uniform crop maturity (Anonymous, 2004).

TABLE 1.3
Drip irrigation under banana before and after drip irrigation

Particulars	Before (surface)	After (Drip : 8 LPH)
Extent of area (ha)	0.2	0.4
Yield (kg)	3200	5804
Water savings (%)	—	29
Maturity	Late	Early (10-15 days)
Uniformity of maturity	No	Yes
No. of harvests	6-8	3-4

CONCLUSIONS

Soil and water conservation (SWC) interventions would help the nature in restoring the flora, though slowly but steadily. They help to moderate the growth process of different vegetation and mitigate the abiotic stresses in the moisture constrained dryland/upland environment. SWC interventions are especially important for high value fruit crops cultivated on dry lands. Different forms of SWC interventions discussed in this chapter are not complete by themselves. A combination of hardware and software treatments can form the holistic approach in improving the hydrologic environment of the system.

An ideal system of SWC interventions moderates the flow hydraulics by moisture conservation *in situ* and storing excess water for mitigating stress situation during dry periods and by draining excess water during high rainfall period to avoid water logging and stagnation. The combination of different soil and water conservation measures, microsite improvement and *in-situ* water harvesting would moderates the hydrological environment of the area devoted to alternate land use system by raising intensive vegetation.

Full benefit of SWC interventions can be realised when they are implemented in participatory mode with the stakeholders.

REFERENCES

Anonymous (2004). Participatory Watershed Management for Sustainable Development in Salayur Watershed, Coimbatore, Tamil Nadu under IWDP, MoRD, Govt. of India, CSWCRTI, Research Centre, Udhagamandalam, Tamil Nadu.

Chandra Prakash, Raizada, A., Samra, J.S. and Sastry, G. (1999). Vegetative Barrier for Resource Conservation. Technical Bulletin, CSWCRTI, Dehradun, Allied Printers, Dehradun.

Das, S.K., Sharma, S., Sharma, K.L., Saharan, N., Nimbole, N.N. and Reddy, Y.V.R. (1993). Landuse options on a semi-arid .Alfisol. American, J. Alternative Agric., **18**(1) : 34-39.

Hudons, N.W. (1987). Soil and Water Conservation in Semi-Arid Areas. FAO Soils Bulletin No. 57, FAO, Rome, 172 p.

Juyal, G.P. and Gupta,. R.K. (1985). Construction of LDPE lined tanks in hills-A case study. Indian J. of Soil Cons., **13**(1) : 10-13.

Katyal, J.C. and Osman, M. (1994). Dryland Horticulture. *In* J.L. Chadda (Ed.) Horticulture Research and Education in India. Indian Council of Agricultural Research, New Delhi.

Khanna, L.S. (1977). Principles and practices of silviculture. Khanna Bandhu, Dehradun, pp. 469.

Lundqvist, J. (1992). Social Forestry and Water Resources: Issues and Studies, Social Forestry Research Digest, ISO, Madras, Vol. 2, 48 p.

Madhu and Khola, O.P.S. (2009). Mixed vegetative barriers for pineapple and grass: An effective soil and water conservation measure for Western Ghats region. Technical Brochure, CSWCRTI, Research Centre, Udhagamandalam, Tamil Nadu, p.12.

Mishra, P.K. and Sharma, S. (1992). Watershed approach in wasteland development. Interaction session on Watershed Development with Focus on Wastelands Development, Regional Centre for Wastelands Development, Report No. 22, pp. 39-44.

Mishra, P.K. and Osman, M. (1995). Pit amendment for higher water retention in degraded alfisols. Proc. All India Seminar on Controlled Irrigation Systems for Horticultural Orchard Crops of Arid Zones. The Institution of Engineers (India), Bhubaneswar, p. 55-59.

Muller, J. and Engelthaler, Z.A. (1985). Bentonite application to sandy soils of Argentina. Report of UNIDO Czechoslovakia Joint Programme for International Cooperation in the field of ceramics, Building material and Non-metallic Minerals based Industries, Pilsen, Czechoslovakia.

Patra, S., Juyal, G.P. and Rathore, A.C. (2013). Annual Report, CSWCRTI, Dehradun, 2013-13, pp. 65-66.

Samra, J.S., Chandra Prakash, Sastry, G. and Raizada, A. (2004). Technical Manual on Contour Trenching for Rehabilitation of Degraded Lands, CSWCRTI, Dehradun.

Sharda, V.N., Sikka, A.K. and Juyal, G.P. (2012). Participatory Integrated Watershed Management – A Field Manual, CSWCRTI, Dehradun.

Sharda, V.N. and Juyal, G.P. (2013). Water Resource Development in the Indian Himalayan Region: Potential and Strategies. *In* : Water Management in the Hill Regions. —Evidence from Field Studies (Eds. Palanisami, K., Sharda, V.N. and Singh, D.V.). Bloomsbury Publishing India Pvt. Ltd., New Delhi.

Sharma, S. and Mishra, P.K. (1994). Strategy for groundwater monitoring and management in small watersheds. International Workshop on Groundwater Monitoring and Recharge in Semi-arid Areas. National Geophysical Research Institute, Hyderabad, SI-163 to 174.

Sharma, S. and Mishra, P.K. (1994). Watershed Management in Dryland Areas – Principles and Practices, p. 497-510. *In* R.P. Singh (ed.) Sustainable Development of Dryland Agriculture in India. Scientific Publishers, Jodhour, India.

Sheng, T.C. (1989). Soil Conservation for Small Farmers in Humid Tropics. FAO Soils Bulletin 60, FAO of UN, Rome.

Singh, G., Venkataraman, C. Sastry, G. and Joshi, B.P. (1990). Manual of Soil and Water Conservation Practices. Oxford and IBH Publishing Co. Pvt. Ltd., New Delhi, India. p. 385.

Sivanappan, R.K. and Aiyasamy, P.K. (1992). Models for Greening the Barren and Wastelands. Experience of NGO's, Government Departments and Farmers Society for Social Forestry Research and Development, Tamil Nadu, Madras, India, 139p.

Weber, F. and Hoskins, M.W. (1983). Soil Conservation Technical Sheets USDA, 95 p.

●●●

2

Adoption of Soil Management Strategies by Fruit Growers —A Big Concern

V.K. SURI, ANIL K. CHOUDHARY AND ANIL KUMAR

Horticulture is one of the oldest fruit plants' cultivation enterprise for human food and nutrition. Today, horticulture is practiced at the individual farm level to a large scale in commercial farms. In corporate farms, the orchard management is done excellently following scientific crop and soil management technologies. But under farmers' level, the fruit crop management is literally done, following mixed blend of traditional, conventional and to some extent, recommended orchard management technology. The farmers apply their knowledge, skills, and also farm technologies to grow fruit crops for human food and non-food uses and for family or social needs. Their orchard operations involves land selection, land development, selection/rearing of planting material in nurseries, planting, fruit husbandry involving crop and soil management, post harvesting and marketing aspects with the aim of improving plant growth, productivity, quality, nutritional value, and biotic and abiotic stress management. Among all these operations, soil management is one of the management aspects of paramount importance.

IMPORTANCE OF SOIL MANAGEMENT ASPECT IN FRUIT CULTIVATION

The establishment of an orchard is a long-term investment which needs very critical planning. The selection of proper location and site, planting system and planting distance, selection of cultivars/nursery plants need again careful planning to ensure maximum production. Crop husbandry with respect to soil management again needs critical attention to harness quality fruit production on sustainable basis. Thus, soil management is one of the major orchard husbandry aspects which need due concern of fruit growers. This chapter, therefore, deals in detail with soil management strategies *vis-à-vis* fruit cultivation with special emphasis to hill and mountain orchard management.

SITE AND LOCATION OF ORCHARDS W.R.T. SOIL FACTORS

Proper selection of site is important for a successful establishment of an orchard. The location should be close to market or close to road so that the transportation is not the problem to fruit growers. The climate of growing area of particular fruit crop should be suitable to grow the selected fruit crop. As for as soil factors are concerned, the selection of site for orchard establishment must be based on following criteria :

- Suitability of soil for the particular fruit crop based on soil fertility and nature of subsoil and soil depth.
- Site must have non problematic soils. If the sites have problematic soils, then their amelioration should be done before orchard establishment.
- Close to water source or adequate water supply round the year with good quality irrigation water.
- Orchard site with proper drainage network.
- In hill and mountain areas, the land must not be too steep. If so, then, site must be workable through formation of contours and terraces before orchard establishment.

LAND AND SOIL MANAGEMENT BEFORE ORCHARD ESTABLISHEMENT

After selection of orchard location and site, the next step is the land devolvement and soil management before orchard establishment. The following field operations can be followed depending upon the site of the orchards :

Contour bunding and terracing

Contour bunding is a proven sustainable land management practice in hilly and mountain eco-systems with marginal sloping lands. Globally, this practice is followed over centuries for a successful orchard and agricultural crop land development. This technology is used traditionally to control soil erosion, promote water retention, and increase crop production with ease of mobility and field operations in the arable land. The contour lines are designed and graded in such a way that flowing water or excess rain water flows through irrigation channels and its speed is slowed down to reduce erosion. Terrace system on other hand, refers to planting in flat land strip, formed across a sloping side of a hilly terrain, lying level along the contours. Terraced fields rise in steps one above the other and help to bring more area into productive use, besides preventing soil erosion. Farmers can use a multi-step process for formation of terraces along contour lines on slopy land (Fig. 2.1) :

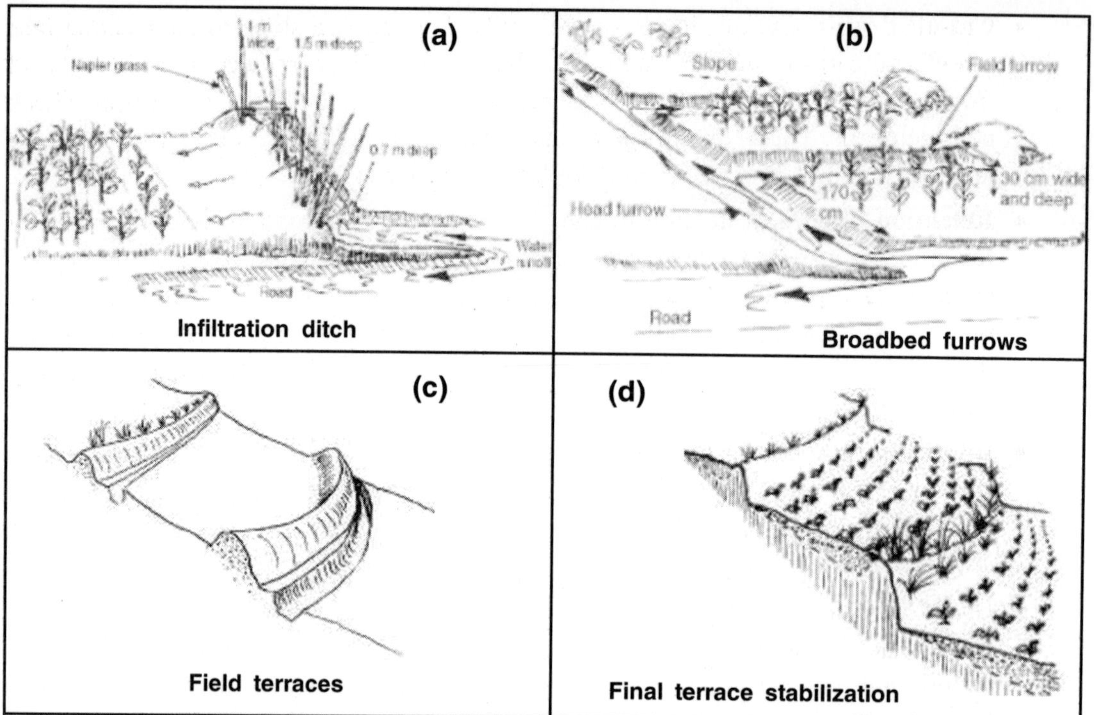

Fig. 2.1. Contour bunding and terracing in hilly and mountain terrains.

- First of all, plant vegetation on the land slope is cut down. All the material is left over to dry and decompose.
- Remaining dry vegetation is rolled on to put along contour lines. Contour lines are the lines that run across hill slope such that the line stays at the same height and does not run uphill or downhill. As contour lines travel across a hillside, they will be close together on the steeper parts of the hill and further apart on the gentle parts of the slope.
- Then, it is strengthened with wooden/tree stumps and stones. This forms rough contour bunds. But, if contour lines are incorrectly established, then they can actually increase the risk of erosion.
- The land between contour bunds is leveled with soil, remaining vegetation litter/ organic matter.
- The process is quite labour intensive and farmers need to regularly check and maintain the bunds to gain stability over time.

Experimental studies have revealed that contouring alone can reduce soil erosion to the extent of about 50% on moderate slopes. For steeper slopes with more than 10% gradient alone, contouring is not enough, thus, other measures also need to be judiciously used to enhance its effectiveness. Therefore, many bio-physical interventions need to be integrated in case of newly made contours and terracing land work.

- **Cut-off drains.** Cut-off drains are generally dug across a slope to intercept surface runoff and carry it safely to an outlet channel in order to protect arable slopy land from uncontrolled runoff, and to divert water from gully heads for erosion check considerations. This is an old age practice in hill and mountain areas. This technology is already under adoption in hill production systems.
- **Retention ditches.** Retention ditches are dug along hill contours. They infuse and retain incoming water runoff and hold it till its seepage in into soil profile. They are an alternative to cut-off drains when there is no nearby waterway to discharge the runoff. This water conservation technology is very effective for the rainfed and dryland areas to harvest rain water in arid and semi-arid agro-ecosystems with hilly terrains (Fig. 2.1a).
- **Infiltration ditches.** Infiltration ditches are one way of harvesting water from various sources of runoff. They consist of a ditch with 05-1.5 m depth dug along contour and upslope from arable field. Water is diverted from catchment into the ditch and blocked at tail end. Water trapped in these ditches seeps into soil profile.
- **Broad-bed and furrow system.** In Broadbed and furrow system, runoff water is diverted into field furrows with 25-30 cm width and 25-30 cm depth. The field furrows are plugged at lower end. In case of high rainfall when one furrow is full; the water backs up into head furrow and flows onto next field furrow and likewise (Fig. 2.1b). In between field furrows, broad beds are framed with appropriate width to grow fruit crops and field crops or vegetables.
- **Hill terraces.** Hill terraces are constructed by digging ditches and heaping the soil to form bunds in the upper sides of ditches. Between ditch and bund, a small ledge prevents the soil from sliding back (Fig. 2.1c and d). Inter terrace spacing depends on land slope as well as soil depth.
- **Vegetative strips.** Grasses or fruit trees are used as vegetative strips to reduce soil erosion and catch surface water runoff as well as excess nutrients. The effect of vegetative strips is multiple with respect to ground cover enhancement, improved soil structure and infiltration as well as reduced soil erosion.

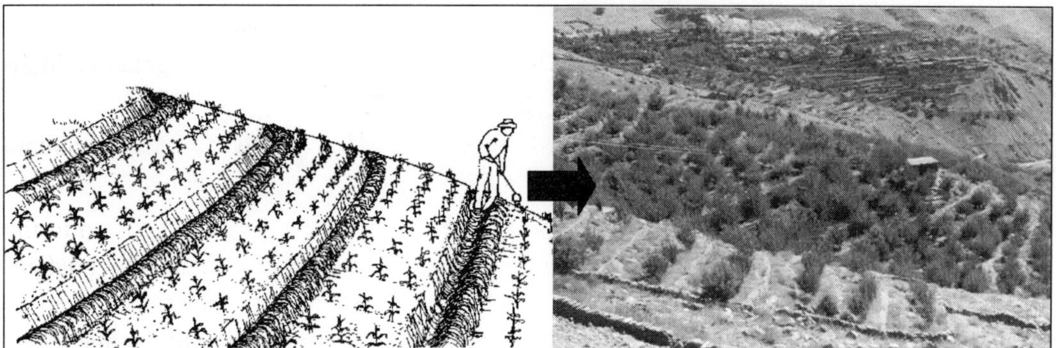

Fig. 2.2. Orchard establishment in contours/terraces in hilly and mountain terrains.

- **Orchard establishment.** Orchard establishment on the hill contours/terraces is a specialized agroforestry land use system where perennial fruit component is grown in combination with field crops within field terraces and pasture crops like Setaria, Napier or any other suitable grasses to both for erosion minimization, terrace stabilization and other ecological and economic considerations (Fig. 2.2).

SLOPING AGRICULTURE LAND TECHNOLOGY (SALT)–AN INNOVATIVE HILL & MOUNTAIN FARMING SYSTEM APPROACH

The sloping agricultural land technology (SALT) is an innovative farming system developed by the Mindanao Baptist Rural Life Centre in the southern Philippines during the 1970's, basically to cate to the production needs of small-scale hill farmers (Choudhary, 2013). SALT is a technology package of soil conservation and food production that integrates several soil conservation measures (Atul and Pratap, 2004; Choudhary *et al.*, 2013a). Generally, SALT technology involves planting field crops and horticultural and tree crops in bands 3-5 m wide between double rows of nitrogen-fixing shrubs and trees planted along the contour. These items minimize soil erosion, retain rain water in soil profile and maintain soil fertility thereby establishing a stable hill ecosystem. The crops provide permanent vegetative cover which aid conservation of both water and soil (Fig. 2.3 and 2.4). This agro-forestry technology has gained wide popularity in Asia; because it is culturally appropriate, economically sound and is designed to limit soil erosion (Atul and Pratap, 2004).

LAYING ORCHARD AND PLANTING PATTERNS

After land development, fields should be thoroughly ploughed, leveled and manured to improve their physic-chemical and microbiological properties before planting operations are taken up.

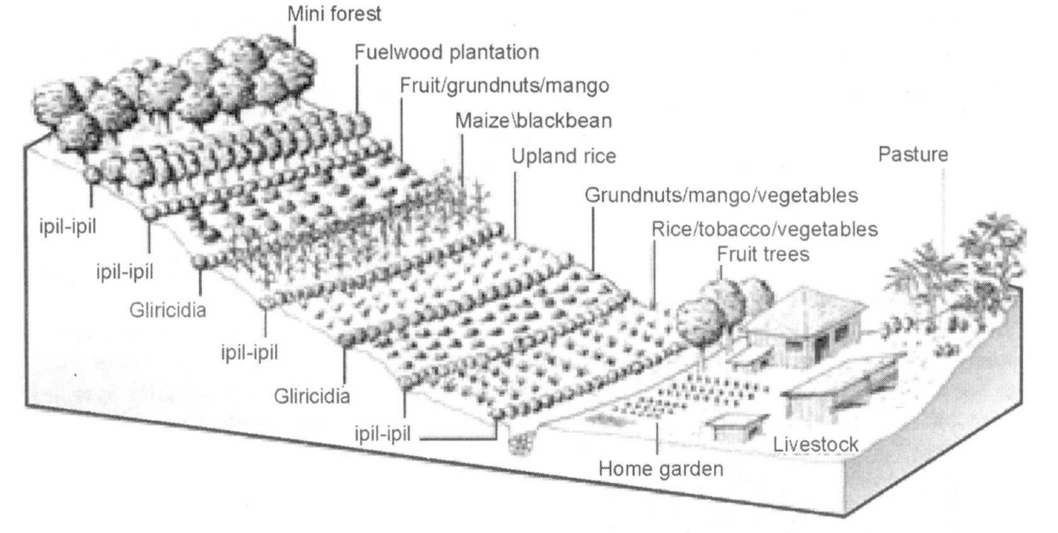

Fig. 2.3. SALT Model–Integrated farming system model : hill and mountain eco-systems.

Fig. 2.4. SALT farming model in NE and NW Himalayan eco-systems, India.

Irrigation channels should be laid along the slope gradients for water economy. Wind break rows of tall trees planted close together around the orchard are essential to resist velocity of wind and reduce moisture evaporation from the soil. Since the wind breaks are very effective in reducing the wind velocity, their presence in strong wind affected areas is of paramount importance. Orchard layout should accommodate maximum number of trees per hectare, adequate space for light interception and plant growth and development and to ensure convenience in orchard cultural practices. Orchard layout can be divided in two broad categories :

(*i*) *Vertical row planting pattern:* In this system, fruit trees are set in a row pattern exactly perpendicular to trees set in their adjacent rows *i.e.* square system and rectangular systems (Fig. 2.5).

(*ii*) *Alternate row planting pattern:* In this planting pattern, fruit trees in the adjacent rows are not exactly vertical; instead plants in even rows are midway between those in odd rows *i.e.* hexagonal, quincunx and triangular systems (Fig. 2.6).

There are many layout systems for orchard planting. In case of hilly areas, another system of planting pattern is followed *i.e.* contour planting. This system is discussed below :

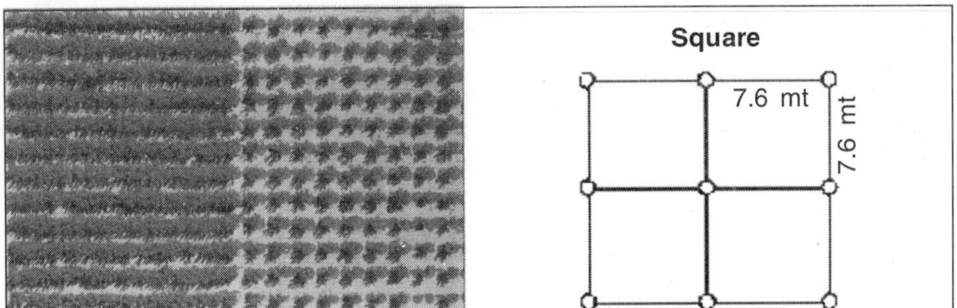

Fig. 2.5. Vertical row planting pattern (Square system).

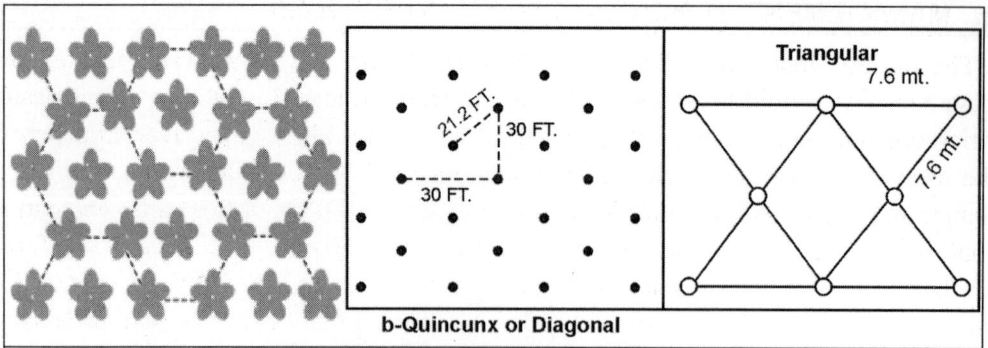

Fig. 2.6. Alternate row planting pattern (Hexagonal, quincunx, and triangular systems)

Contour planting system. *This planting system is* generally followed in hilly areas and the fruit plants are planted along contour across the slope gradient. Thus, it particularly suits to land with undulated hill topography with high danger of erosion and irrigation constraints. Thus, this system minimizes soil erosion, landslides and conserves soil moisture and keeps the slope fit for fruits and plantation crops cultivation. The contour lines are designed and graded in such a way that flowing runoff water or excess rain water flows through irrigation/ drainage channels and its speed is slowed down to reduce erosion (Fig. 2.7A). The contour lines' designing and grading is also discussed earlier in Fig. 2.1, 2.2 and 2.3. Terrace system on the other hand, refers to planting in flat strip of land formed across a sloping side of a hill with leveling along the contours (Fig. 2.7B).

Terraced field plots are made through land work in steps one above the other that help in bringing more area into productive use and also to prevent soil erosion. Contour terrace width varies according to gradient and nature of the slope. If the slope is steep, then we keep the terrace width narrow and *vice-versa* (Fig. 2.7B).

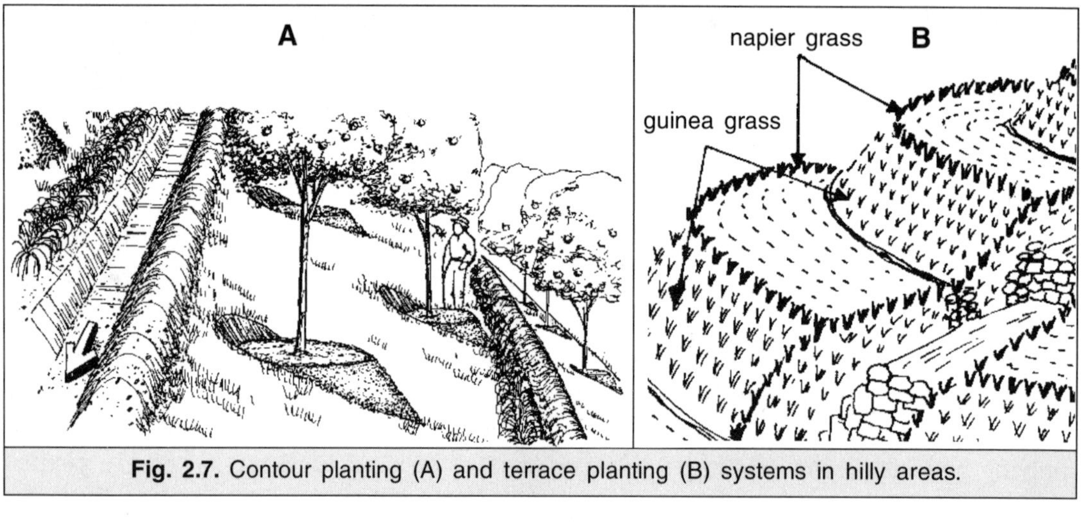

Fig. 2.7. Contour planting (A) and terrace planting (B) systems in hilly areas.

SOIL MANAGEMENT IN HORTICULTURAL NURSERIES

The nursery area should be near the farmers' house as well as water source. Proper drainage facilities should be provided. A medium textured loam/sandy loam soil is desirable with rich organic matter and soil depth of about 15-25 cm. Flat bed and raised nursery beds can be use for nursery raising. Raised bed height should be 10-15 cm with a width of 1 m and length may be according to need and convenience. Two parts of fine earth, one part sand and one part FYM must be incorporated to each bed to improve aeration and soil fertility. Before bed preparation, soil should be drenched with Formalin or 0.3% Copper oxychloride to kill pathogen spores. Time to time, weeding and intercultural operations are necessary for proper growth of nursery plants (Fig. 2.8).

Fig. 2.8. Well managed apple horticultural nursery in Himachal Pradesh.

Planting Methods

After orchard layout, next step is to decide the planting distance for the fruit type to be grown in the selected site. Plating distance depends upon various factors like kind of fruit species, growth habit and plant spread, rainfall and climatic indices of the site, soil type and soil fertility, rootstocks, pruning and training, Irrigation system adopted etc. In general, it is more profitable to plant fruits closer but at appropriate distance. If the trees are too close, there will be plant competition and plants will grow taller needing pruning with difficulty in spraying and harvesting operations. Cultivation also becomes difficult in densely planted orchards. Close planting yields greater per unit area in early life of tree but the same is less in later years.

Planting and Pit Digging

After marking the positions of fruit tree, a pit of about 1 m^3 or of the desired dimensions at the position of the tree marking is then dug out. Pits are then filled with top soil already mixed with well rotten FYM/vermi-compost (Fig. 2.9). Irrigation water is then applied to settle the pit content down properly. The pits are then ready for planting of fruit trees.

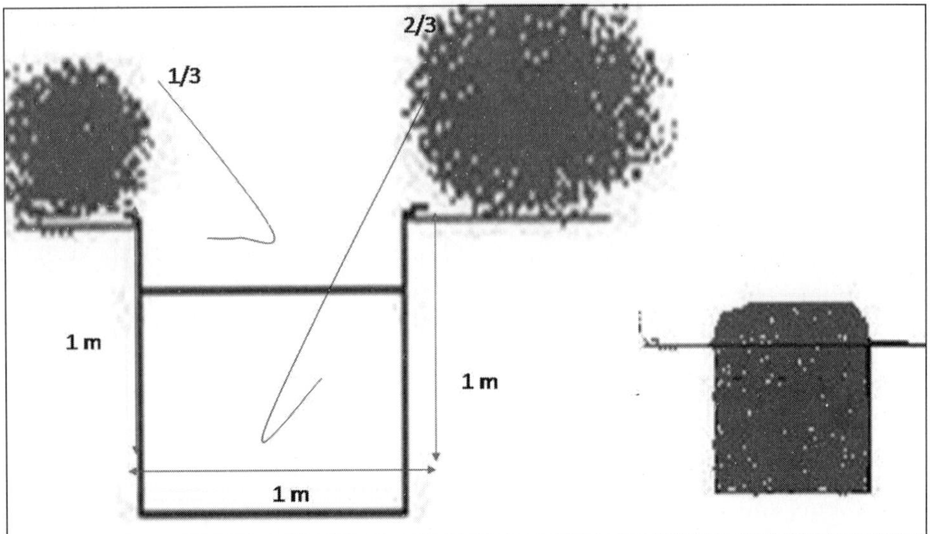

Fig. 2.9. Planting pit making and soil management in pits.

Fruit Transplanting

Fruit trees should be planted in the field as deep as they stood in the nursery. Plants are irrigated just after planting. This helps the roots to establish quickly. A small basin must be

Fig. 2.10. Tree basin formation and straw or plastic mulching in tree basin.

made around the tree for this purpose. During rains, planting basin should be demolished so that water doesn't collect around the basin especially in heavy soils. In rainfed areas, the mulching with straw or plastic mulching in tree basin helps in conserving the moisture (Fig. 2.10).

Orchard Soil Management

It refers to careful management of the orchard soil in such a way that the soil is maintained in a good condition suitable to tree needs with respect to physical condition of the soil, its moisture and nutrient needs. Soil management consists mainly of the conservation of soil, moisture, and fertility and improvement of the physical, chemical and biological properties of soil. These are achieved easily by deep tillage and manuring before every monsoon during the first five years of plantation. It ensures weed management to save moisture and plant nutrients, minimum disturbance to soil and least soil erosion. This involves regular ploughing and removal of weeds or chemical weed management.

Mulching. It is one of the important soil management practice. Crop residues like straw or polythene mulches are spread in the tree basins and in inter spaces between trees (Fig. 2.10). Main objective of mulching is to conserve soil moisture and to control weed growth. Other advantages of mulching are :

- Keeps soil cool in day and warm at night hours
- Reduces surface run-off
- Adds organic matter to the soil
- Prevents soil erosion
- More absorption of rain water
- Reduction in irrigation frequency.

Intercropping and multitier cropping systems

Growing any economic annual crop in alley spaces of the fruit trees in first few years or in unoccupied spaces of a young orchard is called as intercropping. In young fruit orchards especially widely spaced plants, the inter-spaces of the young orchards can be used for intercropping with annual crops like cereals, pulses, vegetables etc. (Badiyala *et al.*, 2012). But, intercrops should not occupy the area where roots of fruit trees are concentrated. Soil fertility should be maintained or improved when intercrops are grown. Water requirements of the intercrops should not compete with main fruit tree needs. The intercropping should be stopped when trees occupy entire orchard space. Certain horticultural plants like coconut and areca-nut are grown for about 50 years in a particular land. It takes nearly 4 to 7 years for the above trees to reach the bearing stage. Adequate alley spaces can be profitably used for raising other crops in multitier system of cropping, thereby enhancing employment opportunities, farm productivity and profitability.

In apple and stone fruit orchards, cole crops like cabbage, cauliflower and broccoli as well as legume vegetable like pea and Fenchbean are the best preferred vegetable crops

Fig. 2.11. Vegetable inter-cropping in young apple orchards in NW Himalayas.

(Fig. 2.11). Black pepper has been trained on mangoes in some places. For growing of intercrops in young orchards, mostly profitable inter-crops are grown till the permanent trees start commercial fruit bearing so that wide spaces in between the young trees are not without proper cultivation. The intercrops should be shallow rooted with short duration with least competition for moisture and nutrients with fruit trees. There is a wide choice from both leguminous and non-leguminous vegetables which come under this class. Vegetables are the best inter-crops in young orchards (Fig. 2.11). Turmeric and ginger are well suited to shady orchard conditions. Among quick growing fruits papayas, bananas and pineapples, encompass their use as intercrops.

TILLAGE MANAGEMENT AND INTERCULTURAL PRACTICES IN ORCHARDS

Soil tillage is one of the important factors affecting soil environment and crop productivity. Tillage methods affect sustainable use of soil resources through its influence on soil properties. Proper use of tillage can improve soil related constrains, while improper tillage may cause a range of undesirable processes involving destruction of soil structure, accelerated erosion, landslides in hills, depletion of organic matter and soil fertility as well. Use of excessive and unnecessary tillage operations is often harmful to soil. Tillage practices modify soil structure by changing its physical properties such as soil bulk density, soil penetration resistance and soil moisture content. All above conditions lead to a favorable environment for crop growth and nutrient use. Tillage including sub-soil tillage should be done thoroughly at this stage, as it cannot be done after planting without disturbing the roots of the fruit trees. Thus, orchard ground floor must have the ability to absorb the rainfall and with high water recharging capacity; this also reduces the soil erosion risks especially in hilly terrains and light soils.

INTEGTARED NUTRIENT MANAGEMENT

Among numerous factors responsible for low productivity of fruit orchards in the country, inefficient nutrient management practices are of prime concern. Application of organics in combination with chemical fertilizers in a judicious manner enhances fruit yields, soil health and resource-use efficiency. Integrated nutrient management (INM) with context to horticulture may be defined as judicious and efficient use of chemical fertilizers (as soil and foliar application) in conjunction with organic manures, orchard residues, green manures, legume inter-cropping systems, use of bio-fertilizers and other locally available nutrient sources for maintenance of soil fertility and for sustaining orchard productivity from all possible sources of plant nutrients in an integrated manner (Choudhary *et al.*, 2013). Chemical fertilizers, organic manures, orchard residues, green manures, legumes and bio-fertilizers are vital components of INM in horticulture :

- **Chemical Fertilizers.** Chemical fertilizers play a prime role in crop production. Use of chemical fertilizers in balanced proportions and recommended amounts keeping in view the fruit tree age and fruiting capacity is the quickest way of boosting fruit production. Chemical fertilizers applied basally in fruit basins alone, are unable to fulfil the fruit nutrient requirements. Thus, foliar application schedule follow-up is another supplemental way to apply macro and micronutrients in fruit farming. Also, chemical fertilizers alone cannot maintain and sustain the long term soil health and fruit productivity because they are unable to supply organic matter and micro-nutrients like organic manures. Hence, the integrated use of inorganic fertilizers with organics is equally important for efficient fruit tree nutrition.
- **Organic Manures.** The organic manures which are mainly prepared from animal's dung/urine and plant residues are referred to as organic manures.
- **Other organics sources.** Some non-conventional materials such as sewage-sludge, urban wastes, press mud, basic slag, fly ash, spent wash and other industrial effluent, etc. could also be the potential sources of plant nutrients.
- **Orchard waste residues.** Orchard residues such as weed flora, leaf litter and crop residues of orchard based intercrops, which are substantially rich source of plant nutrients.
- **Composts.** The process of decomposition of organic wastes is termed as composting and the decomposed material is called as compost.
- **Green Manures.** Green manuring is the easiest and quickest way of supplying readily decomposable organic matter to soils to release plant nutrients.
- **Legumes as inter-crops.** Legumes can form an important component of INM system when intercropped in orchards as crops, as legume fodder or when introduced as green manure crops.
- **Biofertilizers.** These include nitrogen fixers, phosphate solubilizers, mycorrhizal fungi, S and Fe oxidizing bacteria, etc. to be applied in orchard intercrops (Table 2.1).

Overall, integrated nutrient management aims at achieving a harmony in the judicious and efficient use of chemical fertilizers in conjunction with organic manures, farm residues, green manures, legumes, bio-fertilizers and other locally available nutrient sources for sustaining soil health and environment as well as fruit productivity on long term basis.

TABLE 2.1
Some beneficial organisms used as bio-fertilizers in fruits/vegetables

Name of organism	Mode of action	Crops for which used	Method of treatment
Rhizobium	Symbiotic N fixation	Leguminous vegetables/ fodder crops	Seed treatment
Azotobacter	Asymbiotic N fixation	Vegetables	Seed treatment
Azospirillum	Asymbiotic N fixation	Vegetable crops	Seed treatment Soil application
PSM	Phosphorous solubilization	Vegetables, field crops	Seed treatment
AM-fungi	Nutrient (P, Zn and other micronutrients) solubilizer and mobilizer; Water relations	Field crop and fruit crops	Soil application, seed treatment, planting material treatment

(*Source* : Mahajan *et al.,* 2003)

In above background, adoption of mycorrhizal fungi in horticulture, which is an inexpensive and eco-friendly input, is also capable of enhancing both P availability and water-use–efficiency of crops which might also prove very effective in mitigating various stresses (Kumar *et al.*, 2013a; 2013b). Mycorrhizal fungi carry out its functions by expanding surface area of plant root system by 10 to 1000 folds into the soil through their ramifying hyphae, thereby, increasing their exploratory area for harnessing P, other nutrients and water (Suri *et al.*, 2011). Further, mycorrhizal hyphae penetrate the soil pores which are inaccessible to root hairs, thereby, absorbing water which is unavailable to non-mycorrhizal plants (Harrier and Watson, 2003), with greater drought resistance in mycorrhizal plants (Kumar *et al.*, 2013a). The AM symbiosis refers to a mutualistic, symbiotic relationship formed between fungi and living roots of higher plants (Harrier and Watson, 2003). The AM fungi develop an extensive extra-radical hyphal network that grows away from the root, through rhizosphere into surrounding bulk soil matrix (Harrier and Watson, 2003; Suri *et al.*, 2011). Above network makes a significant contribution in improvement of soil structure and water relations (Kumar *et al.*, 2013a). The AM symbiosis can alter plant water relations and responses to drought besides improvement in soil hydraulic properties. Thus, the, role of AM fungi in phosphorus management in field and horticultural crops is crucial especially in Himalayan acid Alfisols (Kumar *et al.*, 2013a; Suri and Choudhary, 2012).

FERTILIZER MANAGEMENT IN FRUIT CROPS

Chemical fertilizers have a prime role as source of plant nutrients in enhancing productivity of horticultural crops. Horticultural crops have tremendous potential for fertilizer use. These are heavy feeders of plant nutrients and a number of crops may absorb 500-1000 kg of N + P_2O_5 + K_2O $ha^{-1}year^{-1}$ or even more under good management conditions (Tandon, 1991). Nutrient uptake by many fruits crops is equal to or higher than that of cereal crops. To replenish removal and to supply sufficient amount of nutrients at each stage of crop growth, adequate rates are needed in fertilizer management schedule of fruit crop (Kemmler and Tandon, 1998). Beside NPK, application of secondary and micro-nutrients is also gaining importance in orchard nutrient management system. Deficiency symptoms of Fe, Zn, Mn, Mo and B have been reported at various places in various crops (Jaggi *et al.*, 2001). Management of micro-nutrients in light soils having less organic matter is very essential. In such soils, application of sufficient organic matter becomes essential to hold micro-nutrients such as Fe, Zn, Mn and Cu by chelation (Chadha and Shikhamany, 1990). Fertilizers also influence quality of horticultural crops, particularly colour, shape, size, taste, shelf life and processing characteristics. Fertilizers also influence the physiology of plant and thereby determine composition of fruit produce and the resistance of fruit plants to various environmental stresses.

For orchard fertilizer management, there is a need for soil and plant analysis for particular growing area. An orchard in commercial bearing can be maintained by adding adequate manuring. However, manurial requirements of fruit trees are governed by many factors like fruit kind, fruit cultivars, climatic conditions and cultural practices followed. Fruit trees will grow and produce on a broad range of soil types. The best yields and longest tree lives occur on loamy, well-drained soil of at least 4 feet depth. Right before bud break is the perfect time to fertilize your fruit trees. If you miss the moment and the trees have begun to bloom, you can still fertilize until June. Don't fertilize in late summer or fall, though, because the new growth put on by the tree can be damaged by frost. Not all fruit trees need fertilizer every year and they don't need it in the same amounts. Too much fertilizer means lots of leaves and shoots, and not a lot of fruit. There are two ways to apply fertilizer to your trees :

- *Spread the fertilizer* in tree basin, rake it in, and then water. Don't start fertilizing next to the trunk. Start a foot from the trunk and spread fertilizer evenly all the way out to the drip line. The *drip line* is at the perimeter of the tree's furthest reaching branches.
- *Digging a series of small holes* is another method of applying fertilizer. Dig the holes six inches down and 12^2-18^2 apart. Start drilling the holes a foot outward from the trunk and continue on to the drip line. Take the fertilizer you've measured out according to the recommended rates and sprinkle a little in each hole until it is used up. This is great for making sure less water soluble nutrients like phosphorus or beneficial mycorrhizae in the fertilizer make it to the tree roots.

FERTILIZER SCHEDULING IN MAJOR FRUIT CROPS

Citrus

Citrus fruits perform well in deep, loose and well aerated soils with pH of 5.5 to 6.2. Usually, citrus growers apply large quantities of farmyard manure (FYM) and organic cakes (Castor, Neem, *Pongamia,* etc.) to improve soil structure and to create favourable conditions for healthy growth. Further, to meet the high demands for nutrients and to maintain plant productivity, organic manures are supplemented with chemical fertilizers. The fertilizer schedule for citrus crops is mentioned in Table 2.2.

TABLE 2.2
Fertilizer scheduling in citrus fruit crops

Age of plant	FYM (kg/tree)	Urea (g/tree)	DAP (g/tree)	MOP (g/tree)
1-3 year	5-15	160-480	35-105	25-75
4-6 year	20-30	640-960	140-205	100-150
7-9 year	35-45	1120-1445	240-310	175-225
10th year and above	50	1600	345	250

***FYM—Farm yard manure; DAP—Diammonium phosphate; MOP—Muriate of Potash.**

Full FYM and P should be applied in January. Whole K and half N fertilizers should be applied 15 days before flowering, while remaining half N should be given after fruit set. Fertilizers should be applied in a circular band at a radial distance of 120 cm from trunk of plant. Deep placement of fertilizers should be avoided because maximum root activity in citrus is found within 15-25 cm. A composite nutrient spray prepared *(Zinc Sulphate 500 g; Copper Sulphate 280 g; Magnesium Sulphate 200 g; Ferrous Sulphate 200 g; Borax 100 g; Manganese Sulphate 200 g; Lime 900 g; Urea 1000 g; Water 100 liters)* and sprayed twice or thrice at 20-25 days interval commencing from appearance of new flush will correct deficiencies and increases the yield and quality of fruits.

Apple

Loamy soils rich in organic matter having a pH of 5.5 to 6.5 with gentle to moderate slope, proper drainage and good aeration are the most suitable for apple plantation. The soil should be free from hard pans and water logged conditions. The presence of lime in soil is good for apple cultivation. Application of FYM and fertilizers start right from planting of an orchard. The first application should be made at the time of filling of pits. The fertilizer dose should depends upon soil fertility, type of soil, kind and age of trees, cultural practices, climate and crop load. The fertilizer schedule recommended for apple crop is as under (Table 2.3) :

TABLE 2.3
Fertilizer scheduling in apple fruit crop

Age of plant	FYM (kg/tree)	Urea (g/tree)	DAP (g/tree)	MOP (g/tree)
1-3 year	5-15	35-105	20-65	50-150
4-6 year	20-30	140-245	85-150	200-350
7-9 year	35-45	315-455	195-280	450-650
10th year and above	50	700	350	700

In matured trees, DAP and MOP should be applied in December. Urea should be applied in 2 split doses. 1st dose should be applied in first fortnight before expected bloom.

Second dose may be applied about 3 weeks after fruit set. Besides, above schedule of nutrients, foliar applications of urea (0.5%), $ZnSO_4$ (0.5%) and $MnSO_4$ (0.5%) at petal fall and $CaCl_2$ (0.5%) at 30-45 days before harvest must be applied for quality fruit size and color (Fig. 2.12).

Fig. 2.12. Well-fertilized young apple orchard with quality fruit size in Himachal Himalayas.

Mango

Mango can be grown on wide variety of soils ranging from alluvial to lateritic with deep (2.0-2.5 m) and well drained soil profile. Suitable pH for mango is 5.5-7.0. Shallow, alkaline, rocky and calcareous soil should be avoided. Good cropping, regular bearing and high fruit quality are prime objectives of bearing trees. For non-bearing trees, liberal doses of NPK are to be applied. During non-bearing stage, N is needed in heavy amounts to support healthy and fast growth. The P is needed for root development and carbohydrate translocation. The K application helps in development of fruit, increases fruit quality and control of fruit drop. During bearing age, NPK fertilization secures sufficient vegetative growth early in the season for next year's growth and to ensure regular bearing. This can be achieved by a heavy dose of nitrogen, a little earlier, than flowering in the '*on year*' to initiate vegetative growth and suppression of bud differentiation so that cropping in the '*on year*' is reduced and the production of vegetative shoots is promoted. The fertilizer schedule recommended for mango crop is as under (Table 2.4) :

TABLE 2.4
Fertilizer scheduling in mango fruit crop

Age of plant	FYM (kg/tree)	Urea (g/tree)	DAP (g/tree)	MOP (g/tree)
1-3 years	5-20	50-100	40-80	100-200
4-6 years	25-50	100-200	80-100	200-400
7-9 years	60-90	200-250	120-160	400-600
10 years and above	100	250	160	600

The FYM and fertilizers are applied in a shallow ring dug around the base of the tree in young orchards. In old bearing orchards, these are broadcasted under plant canopy leaving 30-60 cm from trunk and thoroughly mixed into the soil. The FYM and fertilizers are applied in two split doses. The first dose is given during June-July, consisting of complete N with half K at onset of monsoon. The second dose is applied during September-October with FYM, super phosphate and remaining half of potash. It promotes growth, improves fruit set, yield and quality. The foliar spray is given immediately after the harvest of the crop urea @ 1-2% and another after flowering and at marble stage to improve fruit set and to reduce fruit drop.

Litchi

Fairly deep, well drained loam soils rich in organic matter are best suited for its cultivation. High lime content in soil is also beneficial to its trees. A sandy loam or clay loam with a pH of 5.5 to 7.0 and sufficient soil depth is ideal for litchi cultivation. The acute shortage of NPK seems to stunt all forms of litchi growth, including floral initiation. The fertilizer schedule recommended for litchi is as under (Table 2.5) :

TABLE 5. Fertilizer scheduling in litchi fruit crop

Age of plant	FYM (kg/tree)	Urea (g/tree)	DAP (g/tree)	MOP (g/tree)
1-3 years	5-15	95-285	50-150	25-75
4-6	20-30	380-650	200-300	100-150
7-9	35-45	830-1185	350-450	175-225
10th year and above	50	1490	650	335

The FYM, DAP and MOP should be applied in December, half of urea should be applied in the middle of February and other half in middle of April after fruit set.

Peach

Acidic and saline soils are unfit for peach cultivation, thus, soil pH for peach cultivation should be between 5.8 and 6.8. Deep sandy loam soil is best suited for peach cultivation. Peaches are highly susceptible for water logging and prefer perfect drainage. Peach requires a higher amount of Nitrogen than other temperate fruits. Non-bearing trees should be given 454 g of Nitrogen per tree for every year of age of the tree. The bearing orchard should receive 1100 g each of Nitrogen and potash and double this quantity of phosphorous in inorganic form and 15-25 t FYM ha^{-1}. Whole quantity of FYM along with P and K is given during December-January. Half of N should be given in spring before flowering and the remaining half a month later if irrigation facilities are available. Since active roots of peach tree are present at 0-60 cm radial distance from the trunk and at depth of 0-25 cm the nutrients must be applied in this area followed by mixing up to the depth of 15 cm. The fertilizer schedule recommended for peach is as under (Table 2.6) :

TABLE 6
Fertilizer scheduling in peach orchard

Age of plant	FYM (kg/tree)	Urea (g/tree)	DAP (g/tree)	MOP (g/tree)
1-3 years	5-15	55-170	20-65	65-200
4-6	20-30	225-395	85-150	265-585
7-9	35-45	510-735	195-280	750-1080
10th year and above	50	735	280	1080

The peaches have a high requirement for N and K as compared to P. Thus, DAP and MOP should be applied as single dose in December and half dose of urea should be given 2-3 weeks before normal bloom, while rest half of urea be applied 3 weeks after fruit set.

Pear

It grows well in deep, well drained, fertile, medium textured with relatively more clayay soils. A soil depth of about 180 cm is ideal for proper root growth and fruit production. A neutral pH of 6.0 to 7.5 is desirable because Fe deficiency appears on highly alkaline soils. The fertilizer schedule recommended for pear is as under (Table 2.7) :

TABLE 7
Fertilizer scheduling in pear orchard

Age of plant	FYM (kg/tree)	Urea (g/tree)	DAP (g/tree)	MOP (g/tree)
1-3 years	5-15	35-105	20-65	50-150
4-6	20-30	140-245	85-150	200-350
7-9	35-45	315-455	195-280	450-630
10th year and above	50	525-1050	325-645	730-1500

The DAP and MOP should be applied as single dose in December and half dose of urea should be given 2-3 weeks before normal bloom, while rest half of urea be applied 3 weeks after fruit set.

Plum

Deep sandy loam soils with good drainage are best suited for plum cultivation. Soil depth should be at least 180 cm. The water logged, poorly drained, very shallow soils with excessive salts should be avoided. Application of FYM and fertilizers depends on soil fertility, type of soil, topography, age of tree, cultural practices and crop yield. Full dose of FYM, P and K should be applied during December and January. Half of the nitrogen should be applied in spring before flowering and remaining half a month later. The fertilizer schedule recommended for plum is as under (Table 2.8):

TABLE 2.8
Fertilizer scheduling in plum orchard

Age of plant	FYM (kg/tree)	Urea (g/tree)	DAP (g/tree)	MOP (g/tree)
1-3 years	5-15	55-170	20-65	65-200
4-6	20-30	225-395	85-150	265-585
7-9	35-45	510-735	195-280	750-1080
10th year and above	50	735	280	1080

For a bearing tree of 7 years and above, 40 kg of FYM and 2 kg of calcium ammonium Nitrate (500 g N), 1560 g of single super phosphate (250 g of Phosphorous) and 1,170 g of Muriate of potash (700 g of potash) should be applied. The DAP and MOP should be applied as single dose in December and half dose of urea should be given 2-3 weeks before normal bloom, while rest half of urea be applied 3 weeks after fruit set.

ORGANIC FARMING

Organic farming is another area for specialized orchard farming and soil management. Organic farming includes all types of agricultural production systems, which are environmentally, socially, and economically sound (Suri *et al.*, 2012). A reduction in soil quality as a result of human activities can be defined as soil degradation. Water erosion, wind erosion, chemical degradation, nutrient depletion, loss of organic matter, salinization, acidification, chemical pollution, and deterioration of physical properties are major types of soil degradation (Suri *et al.*, 2012). Specific soil properties that support crop productivity include nutrient reserves, water holding capacity and favourable structure for root growth. Biological attributes are quantity, activity, and diversity of microbial biomass and soil fauna in soils. Organic farming improves the soil quality. The demand for organically grown fruits is also gaining momentum all over the world. It is also increasing in developing countries. Organic farming is another area for specialized crop production for high returns (Choudhary *et al.*, 2010). Overall, organics' application improves the quality of food, soils and environment as well. Organic farming is defined as a production system which largely excludes or completely avoids the use of synthetically compounded pesticides, fertilizers, growth regulators, preservatives and livestock additives (Suri *et al.*, 2012; Yadav *et al.*, 2013). It thus relies upon recycling of crop residues, animal manures, off-farm organic residues and wastes, bio-fertilizers, exploitation of native soil fertility, non-pesticidal methods of pest control and weed management. A suitable definition of organic agriculture is *"It is an ecological production management system that promotes and enhances biodiversity, biological cycles and soil biological activity. It is based on minimal use of off-farm inputs and on management practices that restore, maintain and enhance ecological harmony."*

Benefits of organic farming

Recognizing soil as a dynamic living entity which promotes beneficial biological activities in soil and root zone of plants is central to the theme of organic farming (Choudhary and Shobhna, 2012). Major benefits of organic farming are :

- Conservation of natural resources.
- Prevention of damage to environment.
- Production of clean food.
- Reduction of entry of toxicants in the food chain.
- Prospects of export of organically produced foods.
- Self sustaining production systems dependent largely on farm resources.
- Theoretically, organic agriculture is the most ideal system of farming.

INTEGTARED WATER MANAGEMENT STRATEGIES

Water drainage strategy

Good internal drainage is absolutely essential. Where soil is unsuited to fruit crops, it must be improved prior to planting by tilling or by the incorporation of organic matter. Do not plant trees on any site where water stands for more than an hour following a heavy rain. If a hardpan or poorly drained clay layer is present; poor root development and eventual failure of the planting are likely to result. The trees grown on rocky or sandy soils require more attention and care than trees on deeper soils. On light sandy soils, water and nutrient deficiencies can occur frequently. Well developed drainage system is necessary for horticultural crops.

Irrigation management strategy for horticultural crops

Water is perhaps one of the most critical factors in producing a profitable crop. The irrigation requirement of fruit crops depends upon the crop type and type of soil. The aim of irrigation scheduling is to maintain a continuous high soil moisture level in effective root zone. The amount of water to be applied to the crops depends on different factors including :

- Frequency and duration of drought spells during critical plant growth stages,
- Crop rooting and drought tolerance characteristics,
- Soil water-holding capacity,
- Availability of suitable water source and supply system,
- Cost of water management.

Flood system. This method is followed, where land is flat. The entire area is flooded by letting in water. In this method water is supplied in excessive quantity. This system is commonly practiced in canal or kuhl areas, in wet lands for various crops. The entire area is allowed to saturate with water and the interval between two irrigations is kept fairly long. It also causes stagnation in shallow and ill drained soils.

Basin system. In this system, generally circular basins are made and water let in from the main water channel (Fig. 2.13). This system is useful for loamy soils. The basins size is increased in size as the trees grow. Roots of plants spread much faster than above ground portion. It is, therefore, necessary to irrigate a wide area to supply adequate moisture to the entire root zone.

Fig. 2.13. Well developed basin irrigation system for orchard trees.

Furrow system. Under furrow system of irrigation, entire orchard is ploughed up and divided into furrows (Fig. 2.14). The number of furrows between the rows of trees depends on plant age. When trees are young, a single furrow is sufficient. The size of the furrow varies with the type of soil and slope of the land. Furrow run at right angles to the slope or gradient of land. When land is highly slopy; length of furrow is reduced. Furrows are kept hallow to spread water quickly all over the area. When furrows are deep, water is likely to be absorbed in deeper layers or water intake becomes high. Thus, by adjusting furrow depth, quantity of water to be applied to crops can be controlled (Bana *et al.*, 2013).

Fig. 2.14. Well developed furrow irrigation system.

Ring system. In this system, the water is applied in a ring around the tree. The method is recommended for citrus trees. In this system, water is not allowed to touch the bark of the tree, thereby, reducing the chances of collar rot to which the trees are susceptible. The size of the ring will increase as the trees grow.

Border strip or modified furrow system. In modified furrow system, water is applied from one main channel simultaneously into several furrows. It is let out first in a main feeder channel where it rises up and flows uniformly into all the furrows at the same speed. Initial preparation of land is necessary. Land should be perfectly leveled with gentle slope. Erosion is almost eliminated due to slow flow of water, as water penetrates quickly into deeper layers as horizontal movement is slow.

MICRO-IRRIGATION SYSTEMS

1. Drip/Trickle irrigation

Drip irrigation system considerably saves water and enhances quality and yield of the fruit produce. Drip or trickle irrigation has proved its superiority over other irrigation methods owing to precise and direct application of water in the root zone without wetting entire area. It is most suitable tools for row planted wide spaced fruit crops of high values. It has also ability to supply water soluble fertilizers (Fertigation) in rhizosphere more efficiently. Drip irrigation has comparative advantages over conventional methods (Table 2.9).

TABLE 2.9

Comparative advantages of drip-irrigation over conventional method

Variables	Drip irrigation	Conventional irrigation
Water saving	High : 40-80%	Less due to evaporation run- off, percolation etc.
Irrigation efficiency	80-90%	30-50%
Weed problem	Almost nil	High
Water quality	Saline water up to 3 mmhos/cm can be used	Only normal water can be used
Disease and pest problem	Relatively less	High
Water logging, run-off	Nil	High
Efficiency of fertilizer use	Very high and regulated supply	Heavy loss due to leaching
Range of applicability	In wide range of soil	Not suitable for sandy and undulated type of soil
Yield	20-100% increase	Less compared to drip irrigation

2. Sprinkler irrigation

Sprinkler irrigation is suited for most fruit and field crops. Sprinkler irrigation is a method of applying irrigation water similar to natural rainfall. Water is distributed through a system of pipes usually by pumping. It is then sprayed into the air through sprinklers so that it breaks up into small water drops (0.5-4.0 mm) which fall to the ground. Sprinkler irrigation may be adaptable to any slope, whether uniform or undulating. Sprinklers are best suited to sandy soils with high infiltration rates (25-40 mm/hr) although they are adaptable to most soils.

INTEGTARED WEED MANAGEMENT

Most fruit crops are characterized by low plant density, wide row spacing, and slow initial growth and, as a consequence, by poor competitive ability. The vegetative growth of weeds is so fast those they completely cover fruit plants in initial orchard establishment years. Chemical weed management strategy in fruits is of paramount importance due to cost effectiveness and being labour saver. Many weed management options like manual, mechanical, biological; chemical can also be advocated under IWM in fruit crops :

- **Agronomic practices.** Inter-cropping and mulching, etc.
- **Physical methods.** tillage/harrowing, soil solarisation, etc.
- **Preventive methods.** Quarantine, Legislation, weed transport surveillance etc.
- **Mulching.** Crop residue mulching, leaf mulching, plastic mulching, etc.
- **Biological control.** Use of specific weed enemies–parasites/predators/pathogens.
- **Chemical control.** Best approach is chemical weed control. Selective contact herbicides like Paraquet; and non-selective translocated herbicides like Glyphosate can be effectively used in orchards.

PLASTICULTURE

Protected agriculture or controlled environment agriculture is the modification of the natural environment to achieve optimum plant growth. It is most recent technological intervention of crop production under controlled environments. In these systems, various factors of the environment such as air, temperature, humidity, atmospheric gas composition, nutrient factors etc., are controlled (Choudhary *et al.*, 2012a; 2012b). Many benefits could be obtained by practicing controlled environment agriculture. In case of horticultural nursery raising, use of polyhouses can be done in wet temperate as well as dry temperate areas to some extent.

The main forms of plasticulture include the use of mulches, row covers and poly-tunnels. It has been a common practice to use organic mulches such as straw, dead leaves, etc. to modify the environment to make soil more favourable (weed and moisture control) for plant growth in horticultural crops. Plastic mulches are also used now-a-days for production of high-value crops and pineapple plantations. Plastic mulches with drip irrigation are widely used as irrigation water and fertilizers (Fertigation) could be applied together with the added advantage of reducing cost of production (Rahi and Choudhary, 2011). This is a new area to be explored under horticulture. Plastic sheets can also be used for rain water harvesting in the water storage structures as depicted in Fig. 2.15.

Fig. 2.15. Application of plastics in rainwater harvesting structures for multi-use in integrated farming systems.

CONCLUSIONS

Fruit production mainly depends on certain factors like good quality planting material, soil and agronomic practices, crop management technology with respect to nutrient management, weed management as well as disease and pest management. Among above all, soil management is of prime importance which needs critical attention of fruit growers to harness quality fruit production on sustainable basis. Rapid progress has been made in fruit production systems due to soil and crop management technology. A right land management work to a particular orchard site, right application of soil amendments, development of irrigation and drainage infrastructure, organics addition, planting layout and proper planting, plant nutrition management, intercultural operations, and application of advanced agro-technologies like plasticulture, etc. are some of major soil management practices which need due attention for establishing and maintaining a good orchard in order to harness quality fruit production with least biotic and abiotic stresses to fruit plants in prevalent environmental conditions. Thus, soil and crop management practices should be given paramount importance for higher productivity of fruit crops in holistic manner. As, horticulture creates opportunities for poverty and hunger reduction, regional food security and farm incomes in developing world.

REFERENCES

Atul and Partap, T. (2004). Widening Perspectives on Mountain Agriculture. CSK HPKV, Cen HRD Publication ISBN No : 81-8207-000-7, pp 170.

Badiyala, D., Shekher, J, Sharma, S.K., Singh, R. and Choudhary, A.K. (2012). Agronomic research in hills with special reference to Himachal Pradesh – An overview. *Indian Journal of Agronomy*. 57 (3rd IAC *Special Issue*) : 168-174

Bana, R.S., Vyas, A.K., Choudhary, Anil K., Pooniya, V. and Rana, K.S. (2013). Improved agro-techniques for raising productivity of rainfed areas. *Parsar Doot*. **17**(2) : 1-4.

Chadha, K.L. and Shikhamany, S.D. (1990). Integrated Nutrient Management in Horticultural crops in soil fertility and fertilizer use Vol. IV. Nutrient Management and supply system for sustaining Agriculture in

1990s (Ed. Kumar, V. Shrotriya, G.C. and Kaore S.V.). Indian Farmers Fertiliser Cooperative Limited, New Delhi. pp. 357.

Choudhary, A.K. (2013). Integrated farming systems for hill and mountain ecosystems. *In: International faculty Training to Afghanistan Scientists* held w.e.f. Aug., 2013 to Jan., 2014 at IARI, New Delhi.

Choudhary, A.K. and Shobhna. (2012). Vermi-compost: Its importance and utility in agriculture. *Giriraj.* **34**(42) : 5.

Choudhary, A.K., Rahi, S., Singh, A. and Yadav, D.S. (2010). Effect of vermi-compost and biofertilizers on productivity and profitability in potato in North-Western Himalayas. *Current Advances in Agricultural Sciences.* **2**(1) : 18-21.

Choudhary, A.K., Singh A., Yadav, D.S. and Thakur, S.K. (2013a). Paddy cultivation in terraced fields in hills–A traditional technique for sustainable soil and water conservation. *Indian Farmers' Digest.* **46**(1) : 31-33.

Choudhary, A.K., Thakur, S.K. and Rahi, S. (2012a). Employment Generation and Poverty Alleviation through Protected Cultivation in Himachal Pradesh in North-Western Himalayas: A Case Study. In : '3rd *International Agronomy Congress on Agricultural Diversification, Climate Change Management and Livelihoods'* Nov., 26-30, 2012 at IARI, New Delhi, India. *3rd IAC Extended Summaries, Vol. 3*: pp 1311-1313.

Choudhary, A.K., Thakur, S.K. and Suri, V.K. (2013). Technology transfer model on integrated nutrient management technology for sustainable crop production in high value cash crops and vegetables in North-Western Himalayas. *Communications in Soil Science and Plant Analysis* **44**(11) : 1684-1699.

Choudhary, A.K., Yadav, D.S., Thakur, S.K., Sood, P., Rahi, S. and Arya, K. (2012b). Protected cultivation a profitable venture: Case Study of Mandi district. Pp 19-29. (Eds : Kokate, K.D. *et al.*, 2012). *In*, '*Protected Cultivation in Himachal Pradesh.* ICAR Publication, pp 88,

Harrier, L.A. and C.A. Watson. (2003). The role of arbuscular mycorrhizal fungi in sustainable cropping systems. *Advances in Agronomy* **42** : 185–225.

Jaggi, R.C., Mahajan, A., Choudhary, A.K. and Dogra, R.K. (2001). Plant nutrient deficiency symptoms are partly dependant. *Indian Farmers' Digest.* **34**(9) : 9-11.

Kemmler, G. and Tandon, HLS. (1998). Potassium deficiency and its correction in horticultural crops. IPI, Berne and FDCO, New Delhi. pp 23.

Kumar, A., Choudhary, A.K and Suri, V.K. (2013a). Influence of AM fungi, inorganic phosphorus and irrigation regimes on plant water relations and soil physical properties in okra (*Abelmoschus esculentus* l.) – pea (*Pisum sativum* l.) cropping system in Himalayan acid Alfisol. Journal of Plant Nutrition. (Accepted and in press).

Kumar, A., Suri, V.K. and Choudhary, A.K. (2013b). Mycorrhiza–A magical fungus. *Kisan ki Awaaj.* **4**(7) : 30-32.

Mahajan, A., Choudhary, A.K., Jaggi, R.C., and Dogra, R.K. (2003). Importance of bio-fertilizers in sustainable agriculture. *Farmers' Forum.* **3**(4) :17-19.

Rahi, S. and Choudhary, A.K. (2011). Plastic mulching leads to about thirty percent higher productivity. *Giriraj.* **33**(41) : 5.

Suri, V.K. and Choudhary, A.K. (2012). Fertilizer economy through VA-mycorrhizal fungi under STCR targeted yield model in maize–wheat–maize crop sequence in Himalayan acid Alfisol. *Communications in Soil Science and Plant Analysis*, USA. **43**(21) : 2735-2743.

Suri, V.K., Choudhary, A.K., Chander, G. and Verma, T.S. (2011). Influence of vesicular arbuscular mycorrhizal fungi and applied phosphorus on root colonization in wheat and plant nutrient dynamics in a phosphorus-deficient acid Alfisol of western Himalayas. *Communications in Soil Science and Plant Analysis*, USA. **42**(10) : 1177-1186.

Suri, V.K., Kumar, A. and Choudhary, A.K. (2012). Soil Health Management through Carbon Sequestration under Changing Climatic Scenario. *Progressive Agriculture.* 11 (Conf. issue) : 29-42.

Tandon, H.L.S. (1991). Fertilizer Recommendations for Horticultural crops – A Guidebook. 2nd edition. Fertilizer Development and Consultation Organization, New Delhi. pp. 104.

Yadav, D.S., Thakur, S.K., Sood, P. and Choudhary, A.K. (2013). Assessing training needs of extension functionaries about organic farming in north-western Himalayas. *Journal of Organic Systems, Australia.* **8**(1) : 17-27.

3

Soil Quality Assessment and Enhancement in Fruit Crops

K.L. SHARMA

Indian Agriculture contributes about 18% to gross domestic product and livelihood to 650 million people, whereas horticultural crops contributes 29.65 per cent and plays significant role in countries economy. It has been reported that production of horticultural produce has increased 7 times to that of 1950-51 level. In ranking, India is the second largest producer of fruits with a production of 62.8 mt contributing 11% share in world fruit production. There is about 5.81 mha under fruit crops. Estimates revealed that among the horticulture crops, fruit crops at present recorded a two folds increase and production compared to 1991-92. In the production of some fruit crops *viz*., Mango, banana, papaya, pomegranate, sapota and aonla, India ranks first. The productivity of grapes is the highest in the world. In case of banana and sapota, India has higher national average productivity compared to world average productivity. However, in case of productivity of citrus, mango, apple, guava, pineapple and papaya, India slightly lags behind (Chadha *et al*., 2010). Hughes *et al*. (2012) indicated that fruits and vegetables have a major role in ensuring that the world is not only fed but also nourished. They supply many of the micronutrients that combat malnutrition and contribute to balanced diets required for good human health.

Beside several other factors, appropriate soil management is the prerequisite for successful orchard management and fruit production. Soil plays the role of a growing medium for perennial, deep-rooted fruit trees in a orchard as it provide not only mechanical support for the tree, but also air, water and all the required nutrients for the roots. Soil properties control the availability of soil nutrients and the nutrient uptake by roots. Thus, soil properties have a major influence on the growth of fruit trees and the quality of their fruit (Mishra, 2006). Among the soil quality parameters, the parameters which influence tree growth and fruit production include soil depth, soil pH, EC, soil fertility, organic matter content, soil texture, soil structure, water holding capacity (WHC), bulk density and soil biological conditions.

FOOD PRODUCTION IN INDIA IN RELATION TO WORLD PERSPECTIVE

If we look in the world perspective, food grain yields in India are not low only in comparison with developed countries *viz.*, U.S.A., Canada, Europe, Australia, Japan, but also with those in China, South East Asia, and South America. Similar to food grains, the yields of vegetables in India are about 50% lower than those of the world average, and 60-100% lower than those of China (Pain, 2007; Lal, 2008). Besides several other factors, low yields of crops and cropping systems are also attributed to poor soil fertility and inadequate replenishment of the nutrients. The situation is grimmer in rainfed areas where the crops are poorly nourished or fertilized owing to low soil fertility, low fertilizer use due to poor economic condition of the farmers, monsoon uncertainties, etc., which has resulted in multi-nutrient deficiencies in soils. On the other hand, the changing price policies on fertilizers have made the resource poor farmers of rainfed areas to feed their crops with only certain type of fertilizer, which has resulted in low nutrient use efficiency and profitability due to deficiency/antagonistic relationships of certain essential nutrients. It has been estimated that only 9% of the districts in India use more than 200 kg of $N + P_2O_5 + K_2O$ per hectare (Tiwari *et al.*, 2006). On the other hand, only 32% districts used < 50 kg $N + P_2O_5 + K_2O$ per hectare. Most of the rainfed regions fall in this category. The average fertilizer use during 2008 in the country as a whole was about 117 kg NPK ha^{-1} yr^{-1} (Tiwari, 2008) which was very low when compared with the neighboring countries, like China (277.7 kg ha^{-1}), Japan (290.6 kg ha^{-1}) and Korea Republic (409.7 kg ha^{-1}) as stated earlier. At the same time, the yield levels of some of the crops such as paddy, wheat, maize, etc., in India are significantly lower compared to these countries. Hence, apart from many other reasons, low fertilizer use in India is definitely one of the important causes of low yields not only in field crops but also in Horticultural crops. According to Tiwari (2008), the major challenge ahead to the country is that it needs to have about 30-35 Mt of NPK from different sources to produce 300 Mt food grains to feed its expected population of about 1.4 billion by the end of 2025. Further he added that, if the nutrient removal by other crops like horticulture, vegetables, plantation crops, sugarcane, cotton, oilseeds and potato is considered, nutrient demand curve will touch 40-45 Mt. Further, Katyal (2008) has emphasized that, because of existence of wider gap between nutrient addition and mining by the crops, almost 50% of the Indian soils have reached below the critical limit of plant available zinc in soil. The corresponding deficiency in case of iron is about 25%. The severity of the deficiency has increased because of negligible or no application of organic sources of nutrients. Therefore, to ensure sustainability in production, all possibilities need to be explored to narrow down the nutrient removal use gap in future not only in food grain crops, but also in horticultural and plantation crops.

Hughes *et al.* (2012) reported that, because of the large number of pests and diseases that challenge fruit crops, current crop management practices may often be deleterious to the health of farmers, consumers, and the environment. There is an urgent need to develop technologies that will enhance production of common fruit and vegetable crops in a more ecologically sustainable manner. Patil and Rohidas (2006) comprehensively listed the constraints in cultivation of rainfed fruit crops which include : (*i*) Scanty rainfall with less no. of rainy days, (*ii*) shallow depth, low clay content, high infiltration rates and poor water holding capacity, (*iii*) low organic matter and low fertility, (*iv*) high solar radiation incidence and high wind velocity resulting in high potential evaporation, (*v*) scarce ground water resources and low water table, (*vi*) lack of interest in exploitation of crops, suitable to dry land fruit crops like ber, custard apple, phalsa, aonla etc., (*vii*) higher incidence of pests and diseases.

Mishra *et al.* (2006) have comprehensively emphasized the importance of soil management and related strategies for successful raising of orchards. The key component of their strategies include (*i*) carefully appropriate selection of site from the view point of soil and climatic condition (slope, soil texture) (*ii*) consistently improving and ameliorating soil physical properties to ensure better aeration, water retention and its transmission and movement. This can be achieved by enhancing the use of organics especially organic manures way of other suitable soil amendments, (*iii*) improved drainage by planting the tree along a raised ridge, formation of drainage ditches, or installation drainage pipes, (*iv*) adoption of suitable soil and water conservation practices, (*v*) use of soil amendments for bringing the soil reaction from slightly acidic to neutral.

Thus, from the foregoing information, it is quite evident that among several factors, soil quality plays an important role in successful raising of orchards and improving fruit production. Therefore, much emphasis is needed on selection of suitable site, soil type with good soil quality for planting an orchard. The integration of physical, chemical and biological properties and the resulting level of productivity often are referred to as soil quality. In simplest terms, soil quality is *"the capacity of soil to function"* which reflects the living and dynamic nature of soil. Soil quality should not be limited to soil productivity, but should encompass environmental quality, human and animal health, food safety and quality. Under drought stress situations, the soil with better quality and higher degree of resilience can support the crops in a better manner compared to its counterpart with poor quality and low degree of resilience. This chapter comprehensively deals with the emerging concept of soil quality, soil quality indicators, their monitoring and assessment, management issues, and the importance of soil quality in mitigating the recurring drought effects.

In order to achieve the object of ecological security along with economic and nutrition security, horticultural crops must form a major component of vegetation cover on these lands, because hardy horticultural species (*a*) are adaptable to the extreme conditions of the wasteland areas (*b*) provide a permanent land cover resulting in eco conservation of these lands (*c*) yield nutritionally rich fruits, vegetables and other products (*d*) ensure sustainable income to the people providing insurance against recurrent crop failures and losses (*e*) requires little maintenance and care (*f*) generate employment

LAND DEGRADATION AND SOIL QUALITY DETERIORATION-INDIAN PERSPECTIVE

Land degradation and soil quality deterioration are one of the several reasons of agrarian stagnation both in food grain production and plantation crops and perpetuation of hunger and malnutrition and are major threat to country's food and environmental security. Soil degradation is defined as lowering and losing of soil functions or the decline in the capacity of the soil to produce goods of value to humans and is becoming more and more serious worldwide in recent decades, and poses a threat to agricultural production and terrestrial ecosystem. Soil degradation implies : (*i*) loss in the quality or productivity of soil that is often the result of human activities, such as agriculture practices, deforestation, mining, waste disposal, and chemical spills, (*ii*) changes in soil nutrient status, biota, loss of organic matter, deterioration of soil structure, and toxicity due to accumulations of naturally occurring or anthropogenic materials, (*iii*) loss of agricultural productivity, negative impacts on the environment and economic stability, and exploitation of marginally suitable or virgin land, (*iv*) loss or reduction of soil energy, (*v*) adverse changes in its properties that limit or reduce soils ability to perform agricultural productivity and environmental regulatory capacity and (*vi*) decline in over all soil quality leading to a reduction in biomass productivity and environment moderation capacity.

Depending upon the diversity in rainfall pattern, temperature, parent material, vegetation and relief or topography, the country is taxonomically bestowed with different soil types *viz*., Entisols (80.1 m ha), Inceptisols (95.8 m ha), Alfisols (79.7 m ha), Vertisols (26.3 m ha), Aridisols (14.6), Mollisols (8.0 m ha), Ultisols (0.8 m ha), Oxisols (0.3 m ha) and non-classified soil (23.1 m ha).In the country, total degraded area accounts for 120.7 m ha, of which 73.3 m ha was affected by water erosion, 12.4 m ha by wind erosion, 6.64 m ha by salinity and alkalinity and 5.7 m ha by soil acidity (Anonymous, 2008). The continuous increase in population on one hand results in decrease in farm size and reduction in per capita arable land area on the other. For instance, the per capita land area (ha) of India was 0.15 in 2000 and 0.147 in 2005. It is projected to be 0.136 in 2010, 0.121 in 2020, 0.110 in 2030, 0.105 in 2040 and 0.100 in 2050 (Engelman and Le Roy, 1995). These projections are based on a dubious assumption that there will be no additional conversion of farm land to industrial/ urban and other land uses or to degradation by erosion, salinization, elemental imbalance etc. Yet, soil degradation in India remains to be a serious issue especially that is caused by accelerated erosion by water and wind and by salinization (FAO 1994). The predominant reasons of land degradation and poor soil quality (natural and management linked) can be broadly listed as follows : (*i*) loss of topsoil and organic matter associated with clay size fractions due to water erosion resulting in a 'big robbery in soil fertility', (*ii*) intensive deep tillage and inversion tillage with moldboard and disc plough resulting in (*a*) fast decomposition of crop residues which is catalyzed by high temperature, (*b*) breaking of stable soil aggregates and aggravating the process of oxidation of entrapped organic C and (*c*) disturbance to the habitat of soil micro flora and fauna and loss in microbial diversity, (*iii*) very low levels of fertilizer application and widening of removal-use gap in plant nutrients, (*iv*) mining and other commercial activities such as use of top soil for other than agricultural purpose, (*v*) practicing mono cropping without following any suitable rotation, (*vi*) nutrient imbalance caused due to disproportionate use of primary, secondary and micronutrients, (*vii*) very low use of organic manures such as FYM, compost, vermi-compost and poor recycling of farm based crop residues because of competing demand for animal fodder and domestic fuel, (*viii*) no or low green manuring as it competes with the regular crop for date of sowing and other resources, (*ix*) poor nutrient use efficiency attributing to nutrient losses due to leaching, volatilization and denitrification, (*x*) indiscriminate use of other agricultural inputs such as herbicides, pesticides, fungicides, etc., resulting in poor soil and water quality, (*xi*) water logging, salinity and alkalinity and acid soils.

The two major types of soil degradation are (*i*) natural due to soil forming factors and (*ii*) human-induced due to anthropogenic activities. The adverse alterations in soil properties and processes due to anthropogenic perturbations that end in soil degradation, is a major concern of the modern era. The principal mechanisms of anthropogenic perturbations leading to soil degradation are (*i*) industrial, (*ii*) urban and (*iii*) agricultural. Among these, the agricultural degradative processes are of 3 broad categories *viz.*, physical, chemical and biological (Lal *et al.,* 1989). Lal (2004) reported that physical degradation deteriorates soil physical quality and includes compaction of surface and subsurface soil, erosion of top soil, loss of soil tilth, surface sealing and other characteristics that affect plant roots and water movement. The physical degradation process affects soils mass to volume relationship, air to water relationship, gaseous exchange between soil and the atmosphere, and resistance against disruptive forces of air and water. These processes affect the soils physical, mechanical, rheological (flow) and hydrological properties. The chemical degradation deteriorates soil chemical quality and includes adverse changes in soil reaction or pH, reduction in reserves and availability of plant nutrients, the ability to inactivate toxic compounds and reduce excessive build up of salts in the root zone. Chemical deterioration takes place because of depletion of soluble elements through rainwater leaching, over-cropping and over-grazing, or because of the accumulation of salts precipitated from rising groundwater or irrigation schemes, nutrient depletion due to inadequate fertilizer applications, soil acidification, salinization due to inadequate irrigation and drainage, depletion of organic matter due to fast decomposition and insufficient organic fertilizer and compaction, aggravated by the use of heavy machinery, loss of nutrients, pesticide accumulation and accumulation of toxic elements. Biological degradation adversely affects soil life *i.e.*, biological soil quality attributes and associated biological soil functions. This kind of degradation includes the loss of biodiversity and optimum proportion of different species of soil mesofauna and microorganisms and soil contamination by pathogenic microorganisms (Snakin, *et al.*, 1996). Biological deterioration might be the consequence of depletion of vegetation cover and organic matter content in the soils, denoting a reduction in beneficial soil organisms and soil fauna. The majority of agricultural lands already have depleted levels of soil organic matter (SOM) (McGill *et al.*, 1981; Campbell and Zentner, 1993; Lee and Phillips, 1993).

Changes in soil biological processes can increase soil degradation and adversely affect the quantity and quality of the SOC pool activity and species diversity of soil biota, and increase in relative proportion of soil borne pathogens. Lal and Stewart (1990) have emphasized that these physical, chemical and biological degradative processes of soil not only work independently but also interact with one another and exacerbate the adverse impacts on biomass production and the environment resulting in decline in overall soil quality. Karlen *et al.* (1997) have conclusively reported that the deterioration of soil quality could be a pooled consequence of undesirable physical, chemical and biological process. Hence, assessment of soil quality or health is invaluable in determining the sustainability of land management systems.

Soils of any undisturbed ecology do not face problems such as poor physical conditions, low organic matter, and low fertility. With the intensification of agriculture, without giving any due attention to its physical, chemical and biological attributes, mankind has overexploited the soil-a precious natural resource. The recurring exhaustive use of soil resource has resulted in tremendous decline in its quality. In a natural course, soils have the capacity to restore their original status after combating all kinds of the disturbances over a period of time. This inherent property of soil is termed as resilience. For the same amount of rainfall received, a better soil can support the crops in a better manner compared to its counterpart, which is of poor quality. A soil can be termed as 'better', if it is deep, contains adequate amount of organic matter, have good structure, low bulk density, good infiltration rate and hydraulic conductivity, good fertility/nutrient status, neutral, soil reaction, optimum biological activity, etc. In other words, better soil from the viewpoint of combating drought is one, which can retain more moisture and support good crop for a longer duration during intermittent droughts. Hence, periodical monitoring and assessing soil quality becomes inevitable for higher and sustainable productivity of food and orchard crops. It is important to mention here, that for surface feeding crops, top soil of first 20 cm makes the difference. However, in case of orchard crops, one must look for better soil quality throughout the soil profile. Sustaining soil quality is the most effective method for ensuring sufficient food to support life. Only after the most basic physiological needs of air, water, and food are satisfied, humankind begin to consider their safety and other qualities of life, such as the environment in which they live. But in many countries of the world, including the most developed country like United States of America, soils are being degraded at an alarming rate by wind and water erosion, desertification, and salinization resulting from misuse and improper farming practices. Soils once degraded cannot be renewed over the short term and most definitely not within one generation. Depending upon the kind of soil, it may take about 30 to more than 1000 years to form one inch of topsoil by natural processes.

SOIL QUALITY AND SOIL HEALTH

Much like air or water, the quality of soil has a profound effect on the health and productivity of a given ecosystem and the environments related to it. However, unlike air or water, for which we have legislated quality standards, the definition and quantification of soil quality are complicated by the fact that soil is not directly consumed by humans and animals, as are air and water. According to Doran *et al.* (1999) soil quality is often thought of as an abstract characteristic of soils which cannot be defined because it depends on external factors such as land use and soil management practices, ecosystem and environmental interaction, socioeconomic and political priorities and so on. Perceptions of what constitutes a 'good' soil vary depending on individual priorities for soil function and intended land-use by the stakeholder. The term soil quality and soil health are often used interchangeably in the scientific literature and popular press, with scientists in general preferring soil quality and producers preferring soil health. Some prefer the term soil health because it portrays soil as a living, dynamic organism that functions holistically rather than as an inanimate mixture of sand silt and clay. Others prefer the term soil quality and descriptors of its innate quantifiable physical, chemical, and biological characteristics. Soil health may be considered as the state of the soil at a particular time equivalent to the dynamic soil properties that change in the short term (Singer and Ewing, 2000). In contrast, soil quality may be considered as soil usefulness for a particular purpose over a longer time scale, equivalent to intrinsic or static soil quality. Examples of dynamic soil properties are organic matter content and diversity and number of organisms and microbial constituents or products. In a broad way, soil quality may refer to both permanent soil properties and soil condition.

DEFINING SOIL QUALITY

Historically, soil quality means suitability or limitations of a soil for a particular use. Presently, soil quality has been defined by some scientists as the 'fitness for use' and by others as the "capacity of the soil to function" (Doran and Parkin 1994; Karlen *et al.*, 1997). Seybold *et al.* (1998) defined the soil quality as *'the capacity of a specific kind of soil to function, within natural or managed ecosystem boundaries, to sustain plant and animal productivity, maintain or enhance water and air quality and support human health and habitation.'*

Karlen *et al.* (1997) have emphasized that, important functions performed by the soil include : (*i*) sustaining biological activity, diversity and productivity; (*ii*) regulating and partitioning water and solute flow; (*iii*) filtering, buffering, degrading, immobilizing and detoxifying organic and inorganic materials, including industrial and municipal by-products and atmospheric decomposition; (*iv*)storing and cycling nutrients and other elements within the earth's biosphere; and (*v*) providing support of socioeconomic structures and protection for archaeological treasures associated with human habitation. Collectively, the soil functions are evaluated with respect to their capacity to sustain plant and animal productivity, maintain or enhance water and air quality, and support human health and habitation in determining soil quality (Doran and Parkin, 1994). Quality with respect to soil can be viewed in two ways : (1) as inherent properties of a soil; and (2) as the dynamic nature of soils as influenced by climate and human use and management. With respect to inherent properties, a soil is a result of the factors of soil formation-climate, topography, vegetation, parent material and time. Each soil, therefore, has an innate capacity to function, *e.g.*, some soils will be inherently more productive or will be able to partition water much more effectively than others. The second view of quality relates to the dynamic nature of soils as influenced by human use and management. For example, a farming system that does not protect the surface layer from erosion results in the loss of clay and other finer sized soil particles, organic matter, nutrients and other beneficial properties. In most cases, this eroded soil would be functioning at less than its original potential for agricultural production, and its condition or health would be considered impaired or of a lesser quality. This view of soil quality requires a reference condition for each kind of soil with which changes in soil condition are compared and is currently the focal point for the term 'soil quality'.

SOIL QUALITY INDICATORS AND INDICES

Soils have chemical, biological and physical properties that interact in a complex way to give a soil its quality or capacity to function. Thus, soil quality cannot be measured directly, but must be inferred from measuring changes in its attributes or attributes of the ecosystem, referred to as indicators. Indicators of soil quality should give some measure of the capacity of the soil to function with respect to plant and biological productivity, environmental quality and human and animal health. They should also be used to assess the change in soil function within land use or ecosystem boundaries. Effective indicators (Dalai and Moloney, 2000; Doran and Parkin, 1994; (*i*) respond to change in management practice and provide trends over time (*ii*) integrate soil physical chemical and biological properties and processes (*iii*) be easily measured (*iv*) have drought management expected or threshold values (*v*) have low error associated with measurement (*vi*) be stable in short term to enable measurement (*vii*) not be required to be frequently measured (*viii*) be cost effective (*ix*) have the ability to be aggregated from paddock or site to farm/catchment region (*x*) be mappable in space and time (*xi*) have community acceptance and involvement.

The set of indicators used for assessing soil quality can vary from location to location depending on the kind of land or land use, soil function and the soil forming factors. Soil organic matter is one of the most useful universal indicators of soil quality, because it interacts with other numerous soil components and influences other soil properties. There are also descriptive indicators, which are inherently qualitative and can be used in assessing soil quality. Some of the descriptive indicators include soil crusting/surface sealing, rills, gullies, ripple marks, sand dunes, salt crusting and standing or ponding water. Crop yield (grain or biomass production), plant vigour, rooting patterns, and other aspects of the crop have been used as indicators of soil quality. Crop yield is an important indicator because it gives information about the interacting soil properties of the system as a whole as in a bioassay.

The dimension of scale influences soil quality assessments in both space and time. The area of consideration can be as small as a 'point' on the landscape, a research plot, or as large as a nation or the world. Time frames are important because of the effect of climate, soil moisture conditions, human actions, stage of plant growth, and other factors that give rise to temporal variability in indicator status.

The quantitative assessment of soil quality in true sense cannot be made by using the values of any single indicator individually. Hence, a necessity was felt to integrate the values of predominant indicators to single value index that can be used reliably to monitor and predict the effects of farming systems and management practices on soil productivity, environmental quality, food safety and quality and human and animal health. A valid soil quality index would help interpret data from soil measurements and show whether management and land use are having the desired results for productivity, environmental protection and health (Granatstein and Bezdicek, 1992). Both quantitative and qualitative soil quality indices have been proposed. Qualitative measures of soil quality tend to be more subjective in their measurement, but can be assessed more easily and sometimes being more informative. Based on the farmer's perception of soil quality, a scorecard was developed for assessing soil quality. The scorecard is a farmer-based subjective rating system that places indicators into rating scales of healthy, impaired and unhealthy. The indicators are almost exclusively based on sensory observation *e.g.*, look, feel, and smell. For each indicator a scoring function and realistic baseline and threshold values are established. All indicators affecting a particular soil function are grouped together and assigned a relative weight based on importance. After scoring each indicator, the value is multiplied by the appropriate weight, and an overall soil quality rating is calculated by summing the weighted score for each soil function. Andrews and Carroll (2001) and Andrews *et al.* (2002) used multivariate statistical techniques to determine the minimum data set from physical, chemical and biological indicators that account for at least 85% of the variability in the total data set. They evaluated the efficacy of the chosen minimum data set (MDS) to assess sustainable management by performing multiple regression of each MDS against numerical estimates of environmental and agricultural management sustainability goal *i.e.*, net revenues, peak runoff potential. A soil quality index was developed by transforming the MDS and adding the transformed MDS. Index values exhibited significant differences between management treatments.

ROLE OF SOIL QUALITY IN MITIGATING ILL EFFECTS OF DROUGHT

Availability of moisture in soil is one of the key factors in determining the growth and productivity of field and orchard crops. In rainfed areas, intermittent droughts result in poor growth of field and orchard trees and even sometimes death or mortality under prolonged severe moisture stress conditions. Soil types and their characteristics play predominant role in governing the water retention, release, its movement, and use efficiency and ultimately mitigating the ill effects of drought. The important parameters which influence water retention and release include soil texture (relative proportion of sand, silt and clay), depth of solum layer, soil organic matter content, soil structure, porosity, hydraulic conductivity, aggregate stability, infiltration rate, microbial activity, etc. Soils with deeper solum depth, having higher amount of clay content and organic matter, can retain more water compared to the soils which are shallow, coarse textured and with low organic matter status. The texture of the soil, no doubt cannot be permanently altered. However, some of the soil parameters can be modified by following specific management practices. Based on the experiments conducted over years, it has been established that some of the soil physical parameters relating to water regulation characteristics can be desirably influenced by appropriate tillage practices, water receptive land configurations (soil-water conservation measures), residue recycling, integrated nutrient management practices including green manuring, growing of legumes and other cover crops, balanced fertilization, adoption of appropriate crops and cropping systems. Under any drought situation, soils with improved physical, chemical and biological parameters can effectively protect the field and orchard crops from drought, if other appropriate management practices such as making pits of adequate dimensions (depth and diameter) and ameliorating these pits with suitable soil amendments for planting orchard trees, mulching, maintaining desirable plant/tree population or spacing, growing of relatively drought tolerant genotypes adequate fertilization, weed management, integrated pest management, etc., are followed. Therefore, maintaining and improving soil quality by management practices can help in combating ill effects of drought.

EFFECT OF MANAGEMENT PRACTICES ON SOIL QUALITY INDICATORS AND INDICES

There are numerous reports on the effect of agricultural management practices on soil quality under agricultural crops. Though the analogy of impact of management practices on soil quality indicators and indices in agricultural crops as well as horticultural crops in orchards remains the same, there is not much specific information available in horticulture crops. The salient findings from various short and long term studies pertaining to management practices vs soil quality have been presented here under.

In a long term study conducted in rainfed Alfisol, on integrated nutrient management under reduced and conventional tillage in sorghum-mung bean system, it was reported that irrespective of conventional and reduced tillage, the sole organic treatments out-performed in aggrading the soil quality to the extent of 31.8% over control whereas, the conjunctive nutrient use treatments aggraded the soil quality by 24.2 to 27.2%, and the sole inorganic treatment could aggrade only to the extent of 18.2% over unamended control (Sharma *et al.*, 2008). The extent of percent contribution of the key indicators towards soil quality index (SQI) in this study was: microbial biomass carbon (MBC) (28.5%), available nitrogen ($KMnO_4$ oxidisable) (28.6%), DTPA extractable Zn (25.3%), DTPA extractable Cu (8.6%), hydraulic conductivity (HC) (6.1%) and mean weight diameter of soil aggregates (MWD) (2.9%) (Fig. 3.1). Conjunctive use of organic and inorganic sources of nutrients also proved effective in realizing significantly higher grain yields and sustainability of test crops *viz*., sorghum and mungbean crops.

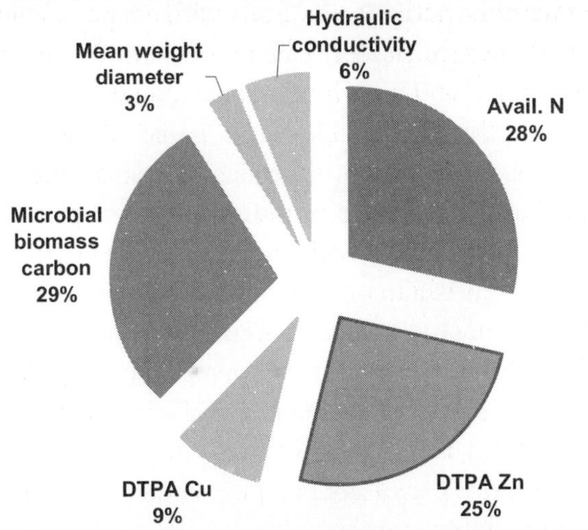

Fig. 3.1. Per cent contribution of different soil quality indicators towards relative soil quality index (RSQI) in Alfisol

Ghuman and Sur (2001) reported that management practices *viz.*, reduced tillage in conjunction with crop residue improved soil properties and subsequent wheat yield on sandy-loam soils in the subtropical climate of northwestern Punjab. Studies on conservation tillage systems, particularly in no-tillage, increased crop residue accumulation on the soil surface. The no-tilled soil had higher values of water soluble C, dehydrogenase, urease, protease, phosphatase and b-glucosidase activities and aggregate stability than tilled soils, but had lower values than the soil under native vegetation. The enzyme activity and aggregate stability showed higher sensitivity to soil management practices than did physical–chemical properties. In this study, it was conclusively brought out that no tillage system was the most effective for improving soil physical and biochemical qualities (Roldan *et al.*, 2005). Beside conservation tillage, positive effect of other soil and nutrient management practices such as green manuring, integrated nutrient management practices, manure application, crop residue recycling, legume based crop rotations, balanced fertilization etc., have been observed on predominant soil quality indicator, overall soil quality indices and sustainability yield indices of crops. Sharma *et al.* (2005) reported that organic carbon in the soil was significantly influenced by long term application (8 years) of crop residues such as sorghum stover and gliricidia @ 2 tons ha^{-1} under minimum and conventional tillage in sorghum-castor rotation in rainfed Alfisols. Further, they reported that increase in nitrogen levels from 0 to 90 kg N ha^{-1} also helped in significantly improving the organic carbon status in these soils over a period of 8 years. From these studies, they concluded that continuous application of organic residues is inevitable to see the significant effect on organic carbon status in soils. Green manuring, which is considered as one of the important practice for improving soil fertility and soil health, is the process of turning a crop into the soil, whether originally intended or not, irrespective of its state of maturity, for the purpose of affecting some agronomic improvement (Mac Rae and Mehuys,

1985). This practice has been found to increase soil N and P availability for the following crop and at the same time, contribute to the conservation of soil organic matter and soil biological, physical and chemical properties (Astier *et al.,* 2006). While studying the management of residues in cropping systems, Smith and Elliot (1990) and Rasmussen and Collins (1991) emphasized the importance of residue application in conserving soil and water and thus ensuring sustainable production, especially in the semi-arid regions where soil and water conservation are of utmost importance. Prasad and Power (1991) emphasized that no one-residue management practice is superior under all conditions. It is important, therefore, to establish under local conditions the beneficial and detrimental consequences associated with a residue management practice before it is propagated among farmers for implementation. It was reported that the application of manure as amendment proved quite effective in improving the soil nutrient status and increasing soil organic C (SOC) levels (Rochette and Gregorich, 1998). The conjunctive use of urea and organics such as loppings of leuceana and gliricidia (1 : 1 ratios on N equivalent basis) had considerable effects on raising the sorghum grain yield to the levels of 16.9 and 17.2 q ha^{-1} respectively and thus revealed that a minimum of 50% N requirement of sorghum can be easily met from farm based organic sources of nutrients (Sharma *et al.,* 2002). Based on a long term study conducted in rainfed Alfisol soils prone to hard setting and crusting, it was observed that integrated nutrient management treatments. *viz*., 2t gliricidia loppings + 20 kg N and 4 t compost + 20 kg N were found to be most effective in increasing the sorghum grain yield by 84.62 and 77.7 per cent over control. However, the highest amount of organic carbon content (0.74%) was recorded in 100% organic treatment (4 t compost + 2 t gliricidia loppings). Some of these options of managing nutrients by using farm based organics can save expenditure on fertilizer and help in improving organic C and overall quality of soil (Sharma *et al*., 2004).

Chaudhury *et al.* (2005) in a study of identifying several biological, chemical, and physical indicators of soil quality concluded that the highest SQI was found in 100% NPK + FYM treatment followed by 100% NPK, 100% NP, 100% N, and control treatment, respectively. A collaborative study coordinated by Mandal (2005) indicated that cultivation without any fertilization (control) or only with N caused a net degradation of soil quality. Cultivation even with application of balanced NPK could hardly maintain such quality at the level where no cultivation was practiced. Only integrated use of organic and inorganic sources of nutrients could aggrade the soil under rice based cropping systems. Manna *et al.* (2005) compared the fertilizer treatments in a long-term study for 30 years in Alfisol (Typic Haplustalf) Ranchi, India. They reported that yield increased with time under NPK + FYM and NPK + lime treatments in wheat crop. While assessing the biological soil health indicators, they found that values of indicators such as soil microbial biomass carbon (SMBC), nitrogen (SMBN) and acid hydrolysable carbohydrates (HCH) were considerably higher in NPK + FYM and NPK + lime as compared to other treatments. Findings of this study suggested that continuous use of NPK + FYM or NPK + lime would sustain yield in a soybean – wheat system without deteriorating soil quality. Soil degradation occurs due to nutrient depletion, soil structure degradation, acidification and sub–optimal addition of organic and inorganic fertilizer to soil. Masto *et al.* (2007) quantified the effects of 10 fertilizer and farm yard manure (FYM) treatments applied for 31 years to a rotation that included maize, pearl millet, wheat and cowpea on an Inceptisol in India. A soil quality index (SQI) based on six soil functions was derived for each treatment using bulk density, water retention, pH, electrical conductivity (EC), plant-available nutrients, soil organic matter (SOM), microbial biomass, soil enzymes and crop yield. SQI varied from 0.55 (unfertilized control) to 0.838 for the combined NPK fertilizer plus manure treatment.

Soil quality indices have been used to compare tillage practices, organic and conventional vegetable production systems and litter management practices in spatially large regions *viz.*, Plains and Hills (Andrews *et al.*, 2003). These varied uses suggest that SQ indices may be applicable not only to different soil types but also to multiple regions and management systems. The impact of long term soil and nutrient management treatments on soil quality using 19 soil chemical, physical and biological indicators has been assessed by Sharma *et al.* (2009) at All India Coordinated Research Project Centers spread across the country and reported that conjunctive nutrient use as well as sole organic nutrient treatments found superior to 100% inorganic nutrient application. Further, he has also suggested the set of key soil quality indicators for each location depending upon the soil type and cropping system. Mandal *et al.* (2001) worked out a crop specific land quality index (LQI) for sorghum [sorghum bicolor (L.) Moench] in semiarid tropics of India. The method developed as LQI is a function of climatic quality index (CQI) and soil quality index (SQI). The LQI was correlated with the actual sorghum yield obtained from benchmarks soils and it was found that LQI bears good agreement with the yield. Doran and Parkin (1994) described a performance based index of soil quality that could be used to provide an evaluation of soil function with regard to the major issues of (*i*) sustainable production, (*ii*) environmental quality (*iii*) human and animal health. They proposed a soil quality index consisting of six elements : SQ = f (SQE1, SQE2, SQE3, SQE4, SQE5, SQE6); Where SQET is the food and fibre production, SQE2 the erosivity, SQE3 the ground water quality, SQE4 the surface water quality, SQE5 the air quality, and SQE6 is the food quality. Awasthi *et al.* (2005) computed integrated soil quality indices in four dominant land uses [forest, upland maize and millet (Bari), irrigated rice (Khet), and grazed systems). Integrated soil quality index (SQI) values varied from 0.17 to 0.69 for different land uses, being highest for undisturbed forest and lowest for irrigated rice. The SQI demonstrated the degradation status of land uses in the following ascending order : irrigated rice > grazed system > forest with free grazing > upland maize and millet > managed forest > grass land > undisturbed forest. The irrigated rice, grazed system, upland maize and millet, and freely grazed forestlands need immediate attention to minimize further deterioration of soil quality in these land uses. From the information presented in this section, it is evident that (*i*) effective soil and nutrient management practices can help in a long way to improve soil quality indicators and overall indices and crop yield sustainability. Hence soil management practices assume great importance in improving soil quality and sustainability. The various soil management practices discussed in this section could be equally effective in improving the quality of orchard soils and enhancing the productivity of fruit crops on sustainable basis. It is only the scale of operation (quantity/method/periodicity) of the practice need to be adjusted in orchard soils.

INFLUENCE OF SOIL AND NUTRIENT MANAGEMENT PRACTICES ON SOIL QUALITY AND ORCHARD PRODUCTIVITY—LESSONS LEARNT

In India, in the past, most of the research studies pertaining to soil quality and sustainability in India was centering around soil testing for only essential plant nutrients, crop response to fertilizer, manures, conjunctive use of fertilizers and organic sources of nutrient on medium and long term basis, computation of optimum levels of fertilizers and manures for recommendation, use of soil amendments to correct acidity, alkalinity, water logging and drainage, protection of top soil through effective soil and water conservation measures etc. Progressively, the research focus shifted with All India Coordinated Projects, where researchers started observing the long term influence of soil and nutrient management treatments on soil parameters through systematic analysis. Many useful results emerged from these studies. Somehow, the prime research focus remained on soil fertility or chemical soil quality indicators except in case of program such as soil structure improvement programs, organic waste recycling, biological nitrogen fixation and few others. In the process of soil quality monitoring, all the three pillars of soil quality (physical, chemical and biological) did not get holistic deal. With advancement in the concept of research on soil quality and sustainability across the world, a paradigm shift in the thinking processes and research programs has come in India also. There are several reports describing the influence of soil and nutrient management treatments such as tillage, residue recycling, and application of organic manures, green manuring and integrated use of organic and inorganic sources of nutrients on soil quality. Some of the key findings of such studies on soil quality conducted in India and abroad in relation to orchard crops are presented in the following section

Hornick (2010) opined that the nutritional quality of field and horticultural crops may be influenced by many factors such as inherent soil composition, climate, the crop and its cultivar, cultural and management practices, and post harvest handling and storage. Roussos and Tassis (2011) reported that citrus is one of the most important fruit tree species in the world, as the fruits are a valuable source of nutrients, vitamins and other antioxidant compounds. They emphasized that among the chemical soil quality parameters, nitrogen plays significant role in initial fruit growth, quality and yield. In orchards where nitrogen level was below optimum, pre-blossom urea applications increased yield significantly (Rabe, 1994). Urea is usually applied as pre- blossom foliar fertilizer, as it is rapidly and efficiently absorbed by leaves and is cost effective. Foliar application of urea has been found to enhance fruit size, peel thickness, juice content, yield as well as total soluble solids (TSS) and acid content in some citrus species (Khan *et al.*, 2009). Beside, nitrogen, micronutrients sprays are also a common practice to rapidly satisfy plant needs (Swietlik, 2002; Tariq *et al.*, 2007). Merwin (2003) reported that hardwood bark mulch and mowed turfgrass reduced N losses, increased soil organic matter, supported excellent fruit yields and improved soil quality substantially in apple orchard. Ghosh *et al.* (1999) observed that VAM (Vesicular arbuscular mycorrhizea) fungi increases P uptake by apple roots. In peach, N has been the main limiting element in Himachal Pradesh, while in plum, maximum fruit yield was recorded when leaf N, P and K contents were 2.89, 0.28 and 0.89% respectively. Akhtar (2005) reported that soil pH exhibited significant and negative correlation with N and P in surface and sub-surface soils and significant and positive correlation with Ca in surface soils. Significant and positive correlation existed between yield of apple fruits and soil N and Ca and leaf N and P concentration. Ghosh *et al.*, (2012) indicated that application of FYM was better than vermi compost with regard to improvement in soil quality which in turn helped in improving plant growth and fruit production in pomegranate. Highest yield (8.1 kg/plant) was recorded in plant where 20 kg FYM along with 400 g N + 100 g P + 300 kg/year was applied and this combination maintained a foliar N/K ratio of 1.3. The treatment also resulted in maximum fruit weight (200 g) with the highest TSS (14.8°B), reducing sugar (12.0%) and vitamin C (12.5 mg/100 ml) content. The fruits from organic fertilizer plant showed maximum storage life with less physiological loss in weight. Singh *et al.* (2012) recorded a maximum yield of 32.15 kg/plant in aonla when treated with FYM along with recommended dose of NPK. Besides improving fruit quality parameters like TSS, total sugars, vitamin C and total phenols, this treatment combination also helped in improving soil quality parameters *viz.*, pH, EC, bulk density, hydraulic conductivity and organic C. Gill *et al.* (2012) reported increased firmness, total sugars and soluble solids in pear fruit with higher dose of K, sprayed thrice with K_2SO_4. Reganold *et al.* (2010) found that the organic farms yielded strawberries with longer shelf life, greater dry matter, and higher antioxidant activity and concentrations of ascorbic acid and phenolic compounds besides soils having more total carbon and nitrogen, greater microbial biomass and activity, and higher concentrations of micronutrients in California. Sofi *et al.* (2012) reported that as essential indicators of soil fertility, soil organic carbon (SOC) and its different labile fractions have an important role in determining soil chemical, physical and biological properties. The results of the study conducted in Shopian district of the Kashmir, Himalayas indicated a significant improvement in the soil organic pools up to 40 years stand age of apple orchards and had a positive effect on the carbon balance. However, in the apple orchard soils in the stand age of 50 years, there was marked depletion in the carbon stocks under the present fertilization practice. At higher altitudes significantly higher levels of particulate soil organic carbon (1.16 g kg^{-1}), labile soil organic carbon (4.94 g kg^{-1}), Walkley and black soil organic carbon (1.44%), microbial biomass carbon (1.30 g kg^{-1}) and total carbon stocks (46.74 Mg ha^{-1}) were recorded; the total soil organic carbon decreased significantly at the lower depths. Other soil organic pools significantly decreased at the 15-30 cm soil depth; particulate soil organic carbon pool showed a reverse trend. In cashew orchards, with a spacing of 6 m × 6 m, highest soil and water conservation efficiency was observed in continuous contour trenches with Stylosanthes seabra + Gliricidia maculata (62.9%) (Maniuannan and Koribanthimath, 2007). It has been emphasized that in case of Apple, the dose of manures and fertilizers should be determined on the basis of soil and leaf analysis. In an orchard of optimal fertility, N, P and K may be applied in the ratio of 70 : 35 : 70/year age of the tree. The dose should be stabilized (700 : 350 : 700 g

N : P : K/tree) after 10 year of age (http://www.fruitipedia.com/apple Malus pumila.html). While reviewing the soil fertility constraints in citrus orchard for Meghalaya region, Srivastava *et al.* (2010) reported that, under low soil pH (pH 4.69-5.55) existing in Khasi mandarin orchards, the problem of Al-toxicity in soil is a prominent problem. The suberisation of walls of roots with corky appearance of root cells which eventually gives root morphologically more thickened in appearance with many dead spot is the major change induced on account of Al-toxicity having soil pH 4.6-4.8. Soil amelioration or fertilization, mainly in the form of organic manure or some Ca based fertilizer holds the prime concern. Fe is considered to be excess and must pose symptoms of toxicity. Srivastava and Singh, (2005) recorded that the symptoms of Fe toxicity was observed in form of excessive defoliation, thereby, plants remain without well developed foliage which result in reduction in fruit yield unless soil pH is amended gradually from 4.5 to about 5.5 in the first phase and to 6.0 in next phase through an expeditiously followed lime application. Zinc is the single most important micronutrient compared to other micronutrients, which not only influence the amount of flowering but their ultimate fruit set intensity. They emphasized the importance of soil microbial population as one of the important soil fertility indicators, since the microbial load of soil is most affected as a result of any management practice. The variations in bacterial count were indicative of soil organic carbon content, which eventually governs the soil microbial buildup and other soil fertility transformations. In this study they also observed that, these bacterial and fungal counts were well correlated with available supply of nutrients like N, P, Zn, etc. muriate of potash, respectively. While suggesting the strategy, of fertility management, they advocated that micronutrients such as zinc sulphate and borax may be used for supplying Zn and B, respectively. Fertilizers should be applied by making full moon terraces. The upper portion of terrace should receive fertilizer to facilitate the movement of fertilizers to the lower half along the gravity. This method lessens the chances of leaching of applied fertilizers.

In order to identify the suitability of different soils for field and orchard crops, NBSS and LUP (2011) has conducted a comprehensive survey of soils of some of the districts of Jammu and Kashmir which includes areas *viz*. Wantipura, Pulwama, Vessut Kakapur, Wahthora and Kamuih. Based on this report, it was observed that soils have medium to high productivity for the crops mainly grown in Kashmir valley. Further, the dominant crops grown in the region are paddy, maize, wheat, mustard, linseed, barley and pulses. Based on suitability criteria, the soils of Pulwama, Vessu, Awantipura and Aharbal soils were found to be suitable for fruit crops also. According to this report, the predominant land constraints in this region were: topography, drainage and inherent soil characteristics which need to be addressed at priority.

ROLE OF ORGANICS IN SOIL MANAGEMENT IN ORCHARD CROPS:

Bloksman (2000) studied the importance of organic soil management in orchard crops and suggested that (*i*) the art of soil management is to decide the optimal strategy for each situation. Organic fruit growers can use some help in recognizing the characteristics of their soil, be aware of the direction they want to develop their orchard and know the technical possibilities to reach that (*ii*) feeding the tree by the roundabout of feeding soil life (*iii*) Fruit growers know especially for fruit you need a well-drained, moist and well-aerated soil. Continuous moderate growth is the basic for fruit production and quality (*iv*) Soil life (bacteria, fungi, worms, etc.) provides such a good soil for fruit. They build a soil structure that drains if wet and sucks up water if dry. Soil life feeds on raw organic matter, root exudates, grass chippings, fallen leaves, prunings, animal manure and decaying weeds. The humification part of soil life produces soil structure and humus. The mineralization part of the soil life converts humus into easily taken up nutrients (*v*) in organic fruit growing, where a good soil structure is a basic condition, we use the roundabout of feeding the tree by feeding soil life. For the extra labour of the application of compost or composted manure you will get in return the labour of soil organisms : a better soil structure, a more regular water household, a better balanced input of all kind of (trace) minerals and less leaching of minerals into ground water. The soil is not a dead substrate just to prevent trees being blown away, but a living organism connected and fed by root exudates of the fruit trees and undergrowth (*vi*) Nutrient balance (*vii*) Mineralization processes provide a steady flux of nitrate from the humus, also when fertilization was long ago. Management of the mineralization process is much more important than the amount of fertilizer. Mitchell (2000) reported that soil organic matter content is frequently identified as a primary attribute of soil quality assessment. SOM influences many soil properties including infiltration rate, bulk density, aggregate stability, cation exchange capacity and biological activity all of which are related to a number of key soil functions. SOM serves as a soil release reservoir for plant macronutrients especially nitrogen and also aids in plant micronutrient nutrition. It facilitates the infiltration of water and air into the soil, increases water retention by the soil and is important for maintaining soil tilth. Over time, increases in SOM can lead to a larger and more diverse population of soil organisms and thus may enhance the biological control of pests and plant diseases. Large quantities of fresh organic matter that are added to the soil, however, may stimulate plant pathogenic organisms and seed and seedling pests such as cabbage maggots and wire worms, which can cause serious losses. Mitchell *et al.* (2000) reported that based on the long interactions, discussions and consensus of researchers and farmers in California, the minimum data sets of physical, chemical and biological properties that can be used as quantitative indicators in soil health assessments could be : physical-bulk density, rooting depth, water infiltration rate, water holding capacity, aggregate stability; chemical-pH, electrical conductivity, cation exchange capacity, organic matter, mineralizable nitrogen, exchangeable potassium, exchangeable calcium; biological-microbial biomass carbon, microbial biomass nitrogen, earthworms, enzymes, disease suppressiveness.

SCOPE FOR HI-TECH HORTICULTURE CROPS—KEY ISSUES

NAAS (2001) categorically emphasized the development in hi-tech production of horticulture crops in the country; there is a need to focus on (*i*) developing high density planting (hdp) (*ii*) developing nutritional need diagnosis and leaf nutrient norms (*iii*) micro-irrigation and fertigation (*iv*) organic farming in horticultural crops. He stressed that leaf analysis, particularly in perennial crops, offers a means of assessing nutritional requirements. Leaf nutrient standards for a large number of crops have been developed and are in use in developed countries. Though, work on these lines is also underway in India but needs to be intensified to ensure economy in fertiliser use through optimised application. To aid practical application, however, there is a need to correlate nutritional composition of leaf with fruit yield on the basis of extensive nutritional surveys of orchards and comprehensive field experiments. Further, he also emphasized the scope and importance of micro-irrigation and fertigation in horticultural crops to make dent in enhancing productivity and quality of fruit crops. Research results have proved that micro-irrigation has its efficacy in saving water and improving yield but its large-scale adoption is limited to a few southern states especially for high value crops. High initial investment on infrastructure has been one of the critical factors. Therefore, NAAS (2001) emphasized on the need to develop a holistic approach to the problem through effective integration of research, development and industry, so that the gap between potential and current status could be bridged. There is need to refine technology, improve design and articulate mechanisms for efficient utilisation of water through micro-irrigation. It is important to mention here that to achieve all this, we have to ensure the ready availability of water-soluble fertilisers for fertigation and to develop guidelines and policies in this regard.

Another important aspect which can benefit the orchardists and consumers is the development of organic farming in horticulture. Organic farming will benefit in two ways. Organic farming will help in producing the safe produce, which will be free from insecticide and pesticide and will also help in improving soil quality in a significant manner. In their quest for higher yields, farmers are using heavy doses of fertilisers, herbicides, fungicides, pesticides, growth regulators. This results in not only increased cost of production but also result in unsafe produce and causes environmental pollution. In recent years, there has been emphasis on organic farming to obtain pesticide residue-free fruits, vegetables, spices and other horticultural commodities. However, scientific technology with sound alternatives is not available. NAAS (2001) advocated that keeping in view the benefits; organic farming offers along with higher returns in international markets for such products, systematic programmes need to be taken up. To ascertain and guarantee the consumer/importer that the produce has been genuinely raised organically, the production has to follow basic organic standards. An Indian Organic Food Production Act, has to be enacted early to safeguard consumers' interest and check exploitation of farmers. Suitable analytical laboratories need to be established, both by the Government of India and state governments, to assess and evaluate the quality of inputs. Research and development on upgrading compost quality, biomass production, biofertiliser and biocontrol measures should be the mandates of all agricultural research centers in the country.

SOIL QUALITY IMPROVEMENT STRATEGIES FOR ENHANCING THE PRODUCTIVITY OF ORCHARD CROPS

The following research, developmental and policy strategies are suggested for effective land care and soil quality improvement for ensuring sustainability on long term basis.

1. Protecting top soil and associated organic matter and nutrients from erosion. It is well accepted connotation that 'Prevention is better than cure'. In order to protect the top soil, organic matter content contained in it and associated essential nutrients, it is of utmost importance that there should be no runoff flow of soil and water out of a given field. If this is controlled successfully, the 'biggest robbery' of clay-organic matter-nutrients is checked. This can be easily achieved, if the already existing technology on soil and water conservation is appropriately used on an extensive scale. The cost for *in-situ* and *ex-situ* practices of SWC (soil water conservation) has been the biggest concern in the past. There is a need to launch 'Land and Soil Resource Awareness Program' (LSRAP) at national level to educate the farmers and orchardists using all possible communication techniques. It is desirable to introduce the importance of soil resource and its care in the text books at school and college levels. The subject at present is dealt apparently along with geography. Farming/orchardist communities too need to be made aware about soil, its erosion, degradation, benefits and losses occurred due to poor soil quality. This can be done through various action learning tools which explain the processes of soil degradation in a simple and understandable manner.

2. Soil and plant test diagnosis based balanced fertilization. Presently about more than 700 soil and plant testing labs are situated in the country. These labs need to be reoriented, restructured and required to be given fresh mandate of assessing the soil quality in its totality including chemical, physical, biological soil quality indicators and water quality and plant analysis. The testing needs to be on intensive scale and recommendations are required to be made on the basis of (*i*) soil test value (*ii*) plant nutrient concentrations in indicator plant part, (*iii*) type of crop/tree (*iv*) previous management history of the field. Special focus is required on site specific nutrient management (SSNM). Soil Health Card (SHC) system needs to be introduced. Soil fertility maps of intensive scale need to be prepared. District soil testing labs need to be renamed as 'District Soil and Plant Care Labs' and required to be well equipped with good equipments and qualified manpower for assessing important soil quality indicators including micronutrients and plant analysis. Keeping in mind the sluggish and inefficient activities of regional soil testing labs of different states, private sector can also be encouraged to take up Soil and Plant Care Programs with a reasonable costs using a analogy of 'Soil and Plant Clinics for Diagnosis and Recommendation' (SPCDR).

3. Care while planting orchards. Besides selecting suitable site using land capability criteria, it is important that, pit size should be of appropriate dimension. These pits can be filled with mixture of soil, FYM, vermicompost, tank silt so that micro-site could be improved for better aeration, water infiltration, root growth and overall establishment of the orchard plant. If required, other suitable amendments such as lime, gypsum, rock phosphate etc., can also be used depending on the soil pH. This approach is referred to as micro site improvement as it influences dynamic indicators of soil quality.

4. Management practices for enhancing organic matter in soil. Enhancing organic matter in soils in semi-arid tropics and tropics is indomitable task. However regular additions of organics without hastening their decomposition process can provide some relief. Management practices such as application of organic manures (composts, FYM, vermi-composts), legume crop based green manuring, tree-leaf based green manuring, crop residue recycling, sheep-goat penning, organic farming, conservation tillage, inclusion of legumes in crop rotation, mulching need to be encouraged (Sharma *et al.*, 2002, 2004). Similar to inorganic fertilizer, provision for incentives for organic manures including green manuring can also be made so that growers should be motivated to take up these practices as inbuilt components of integrated nutrient management (INM) system. These approaches are very effective in improving organic matter in soil and in turn soil quality. In case of orchards, these options can be chosen depending upon the nature of orchard crops/trees. In case of trees, some of these options can be practiced as localized practice in tree basins. It is important to mention here that, compared to field crops, it is easy to enhance organic matter and improve overall soil quality in localized spots/tree basins.

5. Development and promotion of other bio-resources for enhancing microbial diversity and ensuring their availability. In addition to organic manures, there is a huge potential to develop and promote bio-fertilizers and bio-pesticides in large scale. These can play an important role in enhancement of soil fertility and soil biological health. In this way, use of toxic plant protection chemical can also be reduced. In addition to this, there is a need to focus on advance research for enhancing microbial diversity by identifying suitable gene pools. In this case also, more benefit could be accrued in horticulture crops compared to field crops by localized application

6. Ensuring availability of balanced multi-nutrient/customized fertilizers. Fertilizer companies need to produce multi-nutrient fertilizers containing nutrients in a balanced proportion so that illiterate farmers can use these fertilizers without much hassle. While using such products, it is essential that, orchards must be surveyed thoroughly for soil fertility and only needful products containing balanced amount of required nutrients should be used.

7. Enhancing the input use efficiency through precision farming. The present level of use efficiency of fertilizer nutrients, chemicals, water and other inputs is not very satisfactory. Hence, costly inputs go waste to a greater extent and result in monetary loss and environmental (soil and water) pollution. More focus is required to improve input use efficiency. The components required to be focused could be suitable machinery and other precision tools for placement of fertilizers, seeds and other chemicals in appropriate soil moisture zone so that losses could be minimized and efficiency could be increased. This aspect has a great scope in rain fed field crops. However, in case of orchard crops/trees the analogy of 'precision application' can be used in precision spraying and soil application. This will also help in increasing water use efficiency (WUE).

8. Amelioration of problematic soils. History revealed that poor soil quality or degraded soils have taken toll of even great civilizations. No country can afford to let its soils be remaining degraded by virtue of water logging, salinization, alkalinity, erosion etc. Lots of efforts have already gone into the research process in relation to soil amendments. There is a need to ameliorate the soils at extensive scale on regular basis. No matter, how much it costs. Soil amelioration programs should be national programs linked with 'state agricultural departments'. This aspect has great scope in rehabilitating the waste and degraded lands by putting them under valuable multi-purpose tree crops and orchards.

9. Land cover management. The concept of land cover management is still ridiculed at some quarters in India may be because of lack of understanding. The lessons of United States Department of Agriculture (USDA) regarding soil erosion due to wind and water during dusty storms and torrential rains are adequate to understand the concept of land cover management. Covering the land with cover crops such as legumes, natural and pasture grasses, mulches with sparable crop residues will help in protecting the land from the direct hits of high energy raindrops, ill effects of extreme temperatures during summer and winter, reduction in evaporation, enhanced biological activity due to congenial soil habitat conditions, higher C sequestration etc. It has been suggested that mulching could be done either with natural organic waste or plastics. Organic mulch materials include grass mulch, dry leaves, leaf mould, straw, sawdust and green manure. Black plastic mulch is used commonly worldwide but white and aluminized reflective mulches are better for hot climate. Typically, black and clear mulches raise soil temperature above ambient whereas white and aluminized plastics may raise or lower temperature only slightly. Hence, this concept needs to be propagated extensively among the farming community. Development of sod system is another important aspect of protecting the land from erosion and improving overall soil quality in orchards. Mishra (2006) suggested that development of sod system in the orchard has the advantage of cover crops for the entire year. Sod system also helps in protecting the land from erosion, improving organic matter, earthworm activity and improvement in soil structure. This system is well suitable for apple, pear and plum. While selecting the species, it should be ensured that, sod system should be (*i*) able to grow satisfactorily in partial shade, (*ii*) withstand heavy orchard equipment, (*iii*) shallow rooted, (*iv*) suppress nematodes and other insects, and survive drought and winter conditions. Generally, grasses or grass legume mixes are best suited for orchards.

10. Need for structured bodies for land care and soil resource health (BLCSRH). State Soil and Water Conservation departments restrict their activities only up to construction of small check dams, plugging of gullies etc., in common lands. State Soil testing labs are almost sluggish in action, poorly equipped and are with under-qualified manpower. Mostly, no tests are done except for organic C, P and K. State agricultural universities (SAU) only adopt few villages, and consequently, no extensive testing of soil health is done. ICAR institutions also take up few watersheds covering few villages. Then, there will be no one to work for Land Care and Soil Health program at large scale. Hence, **structured bodies at**

Centre and in the States on land care and soil resource health are necessary to effectively coordinate the Land Care and Soil Health Restoration and maintenance programs. It is beyond the capacity of research organizations to take up such giant and extensive task in addition to their regular research mandates. Some of the activities of land care can be linked with National Rural Employment Guarantee Program.

Future research and development outlook in orchard crops in relation to natural resource management

Research focus is needed for developing critical levels of some of the soil quality indicators for which this information is not available for Indian condition. Research experiments should be planned keeping in view three aspects *viz.* soil quality restoration, improvement and maintenance. The subject of soil resilience is still not explored much world over. Systematic research is needed to study soil resilience for diversity of edaphic, climatic and management conditions. Conservation agricultural practices such as conservation tillage, residue recycling, land cover management, appropriate crop rotations have shown the proven benefit to improve soil quality across the world. The quantum of impact may vary depending upon the variations in soil, climate, and duration of the practice and level of overall management of the farms. It would be relevant to study soil quality, resilience and sustainability quantitatively under long term restorative management practices in different crop growing environments. There is a need to develop soil health cards covering important visible and easily understandable indicators so that even illiterate farmers should be able to use them periodically. Beside this, leaf nutrient standards for different horticultural crops should be developed as a ready reckoner and plant indicators.

Some other important points needs to be given research and development emphasis include (*i*) selection of suitable combination of various land uses for optimum utilization of land, water and natural energy sources, (*ii*) selection of suitable fruit, vegetable, fodder and fuel trees with suitable root system desirable for binding the soil and by way of capitalizing land capability – crop/tree suitability concept, (*iii*) to develop technology for fast rehabilitation of degraded waste and jhum lands and to put them under valuable tree species including fruit crops, (*iv*) more research and development focus on water harvesting and its recycling in orchard raising especially in hilly regions (*v*) more research and development focus on micro irrigation and fertigation.

REFERENCES

Akhtar, F. (2005). Studies on relationship between fruit yield and quality with soil and leaf nutrient content in apple orchards of Zangier block of district Baramulla Kashmir. Sher-e-Kashmir University of Agricultural Sciences (Srinagar).

Andrews, S.S. and Carroll, C.R. (2001). Designing a decision tool for sustainable agroecosystem management: Soil quality assessment of a poultry litter management case study. *Ecol. Applic.* **11**(6) : 1573-1585.

Andrews, S.S., Karlen, D.L. and Mitchell, J.P. (2002). *A comparison of soil quality indexing methods for vegetable production systems in Northern California.* Agric. Ecosyst. Environ. **90** : 25-45.

Andrews, S.S., Flora, C.B., Mitchell, J.P. and Karlen, D.L. (2003). Growers' perceptions and acceptance of soil quality indices. *Geoderma* **114** : 187-213.

Astier, M., Maass, J.M., Etchevers-Barra, J.D., Pena, J.J. and Gonzalez, F.D. (2006). Short-term green manure and tillage management effects on maize yield and soil quality in an Andisol. *Soil and Tillage Research*, **88** : 153-159.

Awasthi, K.D., Singh, B.R. and Sitaula, B.K. (2005). Profile carbon and nutrient levels and Management effect on soil quality indicators in the Mardi watershed of Nepal. Acta Agriculture Scandinavica Section B- Soil and Plant. **55** : 192-204.

Bloksman, J. (2000). **Soil management in organic fruit growing. Symposium organic fruit hri-Emra-Adas-Wye College, Ashford (Kent) 16-17. Pagina.**

Campbell, C.A. and Zentner, R.P. (1993). Soil organic matter as influenced by crop rotations and fertilisation. *Soil Science Society of America Journal.* **57** : 1034-1040.

Chadka, K.L., Singh, A.K. and Patel. V.B. (2010). Recent Initiatives in Horticulture Westville Publishing House. Pp-736.

Chaudhury, J., Mandal, U.K., Sharma, K.L., Ghosh, H. and Biswapati Mandal. (2005). Assessing soil quality under long-term rice based cropping system. *Communications in Soil Science and Plant Analysis.* **36**(9/10) : 1141-1161.

Dalai, R.C. and Moloney, D. (2000). Sustainability indicators of soil health and biodiversity. In 'Management for sustainable ecosystems'. (Eds P Hale, A Petrie, D Moloney, P Sattler) (Centre for Conservation Biology, The University of Queensland : Brisbane).

Doran, J.W. and D.M. Linn. (1994). Microbial ecology of conservation management systems. *In* Advances in Soil Science. Soil Biology : Effects on Soil Quality. J.L. Hatfield, and B.A. Stewart (eds.). Lewis Publishers, CRC Press, Boca Raton, FL, pp. 1-27.

Doran, J.W., Jones, A.J., Arshad, M.A., Gilley, J.E. (1999). Determinants of soil quality and health. In : Pankhurst CE, Doube BM, Gupta VVSR (eds) Biological indicators of soil health. CAB International, Oxon.

Engelman, R. and P. Le. Roy. (1995). *Conserving Land : Population and Sustainable Food Production.* Population and Environment Programme, Population Action International, Washington, DC.

Ghosh, S P. (1998). Deciduous Fruit Production In India. In : Deciduous fruit production in Asia and The Pacific held in Bangkok, Thailand, 2-4 June.

Ghosh, S.N. Bera, B., Roy, S. and Kundu, A. (2012). Integrated nutrient management in pomegranate grown in laterite soil. *Indian Journal of Horticulture.* **69**(3) : 333-337.

Ghuman, B.S., Sur, H.S. (2001). Tillage and residue management effects on soil properties and yields of rainfed maize and wheat in a subhumid subtropical climate. *Soil Tillage Res.* **58** : 1-10.

Gill, P.P.S., Ganaie, M.Y., Dhillon W.S. and Nav Prem Singh. (2012). Effect of foliar sprays of potassium on fruit size and quality of 'Patharnakh' pear. Indian J. Hort. **69**(4) : 512-516.

Granatstein, D., and D.F. Bezdicek. (1992). The need for a soil quality index: Local and regional perspectives. *Am. J. Alt. Agric.* **7** : 12-16.

Hornick, S.B. (2010). Nutritional quality of crops as affected by management practices. http://www.infrc.or.jp/english/KNF Data Base Web/PDF.

Hughes, J.A., González. A., Keatinge, J.D.H., Wyckhuys, K., Alvarez. E. and Luther. K. (2012). Eco-Efficient Research to Provide Safe, Profitable, and Environmentally Sustainable Production of Fruits and Vegetables International Center for Tropical Agriculture (CIAT), Cali, Colombia.

Karlen, D.L., Mausbach, M.J., Doran, J.W., Cline, R.G., Harris, R.F. and Schuman. G.E. (1997). Soil quality: A concept, definition, and framework for evaluation. *Soil Sci. Soc. Am. J.* **61** : 4-10.

Katyal, J.C. (2008). Presidential Address. *Journal of the Indian Society of Soil Science.* **56** (4) : 322-324.

Khan, A.S., Malik. A.U., Perez. M.A., Saleem. B.A., Rajwana. I.A., Shaheen, T. and Anwar. R. (2009). Foliar application of low-biuret urea and fruit canopy position in the tree influence the leaf nitrogen status and physico-chemical characteristics of Kinnow mandarin (Citrus reticulata Blanco). Pak. J. Bot. **41** : 73-85.

Lal, R. (2004). Carbon sequestration in Dryland Agriculture. Crop Science Society and Am. Soc. of Agron. CSSA Special Publication No. **32** : 315 -334.

Lal, R. (2008). Soils and India's Food Security. Journal of the Indian Society of Soil Science, **56**(2) : 129-138.

Lal, R., and Stewart. B.A. (1990). Soil degradation: Advances in soil science Springer-Verlag, New York.

Lal, R., Hall, G.F., Miller, P. (1989). Soil degradation: I. Basic processes. Land Degradation and Rehabilitation. **1** : 51-69.

Lee, J.J., Phillips, D.L., 1993. The effect of trends in tillage practices on erosion and carbon content of soils in the US CornBelt. Water Air Soil Pollut. **70** : 389-401.

MacRae, R.G. and G.R. Mehuys. (1985). Effect of Green Manuring on Physical Properties of Temperate-Area Soils. Adv. in Soil Science, Vol.3, Springer-Verlag, New York, pp. 71-90

Mandal, D.K., Mandal, C. and Velayutham, M. (2001). Development of a land quality index for sorghum in Indian semi-arid tropics (SAT). Agricultural Systems **70** : 335-350.

Maniuannan, S., Koribanthimath, V.S. (2007). Soil and Water Conservation Measures for Sustainable Production of Cashew. ICAR Report.

Manna, M.C., Swarup, A. and Wajari, R.H. (2005). Long-term effect of fertilizer and manure application on soil organic carbon storage, soil quality and yield sustainability under sub-humid and semi-arid tropical India. Field Crops Res. **93** : 264-280.

Masto, R.E., Chhonkar, P.K., Dhyan Singh, and Patra, A.K. (2007). Soil quality response to long-term nutrient and crop management on a semi-arid Inceptisol. Agriculture, Ecosystems and Environment **118** : 130-142.

McGill, W.B., Campbell, C.A., Dormaar, J. F., Paul, E.A., and Anderson, D.W. (1981). Soil organic matter losses. Agricultural land: Our disappearing heritage. A Symposium. In: *Proceedings of the 18th Annual Alberts Soil Science Workshop,* Edmonton, 24-25 February 1981. Alberta Soil and Feed Testing Lab., Edmonton, pp. 72-133.

McGill, W.B., Campbell, C.A., Dormaar, J.F., Paul, E.A. and Anderson D.W., 1981. Soil organic matter losses. Agricultural land: Our disappearing heritage. A Symposium. In: Proceedings of the 18th Annual Alberta Soil Science Workshop, Edmonton, 24-25 February 1981. Alberta Soil and Feed Testing Lab., Edmonton, pp. 72-133.

Merwin, I.A. (2003). Orchard groundcover management: long-term impacts on fruit trees, soil fertility, and water quality Proc. N. Engl. Fruit Meetings 2002-2003., **108-109** : 59-65.

Mishra B. (2006). **Management of Orchard Soil : New Approaches. In : Integrated crop management in subtropical fruits (eds) Tiwari, J.P., Srivastava Ranjan. Published by Geeta Somani Agrotech Publishing Academy, Udaipur. P. 62-68.**

Mitchell, J.P., Lanini, W.T., Temple, S.R., Brostrom, P.N., Herrero, E.V., Miyao, E.M.,Prater, T.S., Fouche, B., Mullen, R.J. and Hembree, K.J., (2000). Conservation tillage initiatives in California. In: Proceedings of Conservation Tillage 2000: Con-servation tillage Success Stories from around the U.S., Parlier, CA, USA, 10–11February 2000, p. 83.

NAAS (2001). Hi Tech Hiorticulture in India Policy Paper 13 October 2001 Pp 9.

Pain, A.K. (2007). Indian agriculture: The present scenario. *Agrarian Crisis in India: An Overview* (A.K. Pain, Ed.), pp. 1-11. The ICFAI Press, Hyderabad, India.

Patil, M.B. and Rohidas, S.B. (2006). Major constraints in cultivation of rainfed fruit crops. In: Production and management of rainfed fruit crops. (Eds: M.B. Sontakke). Agrotech Publishing academy, Udaipur – 313002.

Prasad, R. and Power, J.F. (1991). Crop residue management. Advances in Soil Science **15** : 205-251.

Rabe, E. (1994). Yield benefits associated with pre-blossom low-biuret urea sprays on Citruss. pp. J. Hort. Sci. **69** : 495-500.

Rasmussen, P.E. and Collins, H.P. (1991). Long-term impacts of tillage, fertilizer, and crop residue on soil organic matter in temperate semiarid regions. Advances in Agronomy **45** : 93-134.

Reganold, J.P., Andrews, P.K., Reeve, J.R., Carpenter-Boggs, L. and Schadt, C.W. (2010). Fruit and Soil Quality of Organic and Conventional Strawberry Agroecosystems. PLoS ONE **5**(9) : e12346. doi:10.1371/ journal.pone. 0012346.

Rochette, P. and Gregorich, E.G. (1998). Dynamics of soil microbial biomass C, soluble organic C and CO_2 evolution after three years of manure application. *Canadian Journal of Soil Science.* **78** : 283-290.

Roldan A., Salinas-García J.R., Alguacil M.M., Caravaca F. (2005). Changes in soil enzyme activity, fertility, aggregation and C sequestration mediated by conservation tillage practices and water regime in a maize field. *Applied Soil Ecology*, **30** : 11-20.

Roussos, P.A. and Tassis, A. (2011). Effects of girdling, nitrogen, zinc and auxin foliar spray applications on mandarin fruit "Nova" quality characteristics. Emir. J. Food Agric. 2011. **23**(5) : 431-439.

Seybold, C.A., Mausbach, M. J. Karlen, D.L., and H.H. Rogers. (1998). Quantification of soil quality. p. 387-404. In : R. Lal, J.M. Kimble, R.F. Follett, and B.A. Stewart (eds.). Soil Processes and the Carbon Cycle. CRC Press, Boca Raton.

Sharma, K.L. Srinivas, K., Das, S.K., Vittal, K.P.R. and Kusuma Grace, J. (2002). Conjunctive Use Of Inorganic and Organic Sources of Nitrogen for Higher Yield of Sorghum in Dryland Alfisol. *Indian Journal of Dryland Agric. Res. & Dev.* **17**(2) : 79-88.

Sharma, K.L., Kusuma Grace, J., Srinivas, K., Ramakrishna, Y.S., Korwar, G.R., Maruthi Sankar, G., Uttam Kumar Mandal, Ramesh, V., Hima Bindu, V., Madhavi, M., and Pravin N. Gajbhiye. (2009). Influence of Tillage and Nutrient Sources on Yield Sustainability and Soil Quality under Sorghum-Mung Bean System in Rainfed Semi-Arid Tropics. *Communications in Soil Science and Plant Analysis.* **40**(15) : 2579-2602.

Sharma, K.L. Kusuma Grace, J. Uttam Kumar Mandal, Pravin N. Gajbhiye, Srinivas, K. Korwar, G.R. Ramesh, V., Kausalya Ramachandran, and S.K. Yadav. (2008). Evaluation of long-term soil management practices using key indicators and soil quality indices in a semi-arid tropical Alfisol. *Australian Journal of Soil Research.* **46** : 368-377.

Sharma, K.L., Srinivas, K., Mandal, U.K., Vittal, K.P.R., Kusuma Grace, J., and Maruthi Sankar, G. (2004). Integrated Nutrient Management Strategies for Sorghum and Green gram in Semi arid Tropical Alfisols. *Indian Journal of Dryland Agricultural Research and Development.* **19**(1) : 13-23.

Singer, M.J. and S.A. Ewing. (2000). Soil quality. In Handbook of Soil Science. M.E. Sumner (eds.). CRC Press Inc., Boca Raton, FL. P. G 271-298.

Singh, A.K., Singh, S. and Appa Rao, V.V. (2012). Influence of organic and inorganic nutrient sources on soil properties and quality of aonla in hot semi-arid ecosystem. *Indian Journal of Horticulture.* **69**(1) : 50-54.

Smith J.L., and Elliott, L.F. (1990). Tillage and residue management effects on soil organic matter dynamics in semi arid regions. *Advances in Soil Science,* **13** : 69-85.

Snakin, V.V., Krechetov, P.P., Kuzovnikova, T.A., Alyabina, I.O., Gurov, A.F., and Stepichev, A.V. (1996). The system of assessment of soil degradation. *Soil Technology* **8** : 331-343.

Sofi, J.A., Rattan, R.K., Datta, S.P. (2012). Soil Organic Carbon Pools in the Apple Orchards of Shopian District of Jammu and Kashmir *Journal of the Indian Society of Soil Science* Year : 2012, Volume : 60, Issue : 3

Srivastava, A.K., Sing, I.P. and A.K. Das. (2010). Citrus Production Constraints In Meghalaya: Issues And Strategies.

Swietlik, D. (2002). Zinc nutrition of fruit trees by foliar sprays. Acta Hortic. **594** : 123-129.

Tariq, M., M. Sarif, Z. Shah and R. Khan. (2007). Effect of foliar application of micronutrients on the yield and quality of sweet orange (Citrus sinensis L.). Pak. J. Biol. Sci. **10** : 1823-1828.

Tiwari, K.N. (2008). Future of plant nutrition research in India. *Journal of the Indian Society of Soil Science* **56** : 327-336.

Tiwari, K.N., Sharma, S.K. Singh, V.K., Dwivedi, B.S. and Shukla, A.K. (2006). Site Specific Nutrient Management for Increasing Crop Productivity In India. PDCSR (ICAR) – PPIC, India Programme, pp 92.

Uttam Kumar Mandal., Rao, K.V., Mishra, P.K., Vittal, K.P.R., Sharma, K.L., Narsimlu, B., and Venkanna, K. (2005). Soil infiltration, runoff and sediment yield from a shallow soil with varied stone cover and intensity of rain. European Journal of Soil Science, **56**(4) : 435-444.

● ● ●

4

Statistical Procedures for Analysis of Data of Horticultural Crops

G.R. MARUTHI SANKAR

In the field studies with different horticultural species, different statistical methodologies are adopted for planting and management of species, collection of data of soil and plant parameters, statistical modeling and analysis of data over years. In this paper, different procedures are described about the statistical analysis of data and modeling procedures for assessing the growth of species over a time period. Statistical analysis of horticultural data comprises of different growth parameters like tree height, girth, canopy spread in different directions, fruit yield, fruit weight, fruit quality parameters, apart from soil physical and chemical parameters etc. For testing difference between input treatments, micro-site environments etc., horticultural field experiments are conducted based on suitable statistical designs like completely randomized, randomized block, split-plot, latin square design, augmented designs etc. Randomization is essential since it assigns equal probability for all treatments while getting allotted to experimental plots. Replication is essential for deriving an estimate of error, otherwise no statistical test is possible either for comparing or selecting treatments. Local control is the principle of dividing a field into homogeneous blocks, such that plots within a block are homogeneous and between blocks are heterogeneous. A suitable estimate of experimental error is derived for testing treatment differences based on F-test which is computed as a ratio of treatment mean sum of squares and error mean sum of squares. The hypothesis of no differences (or null hypothesis) is rejected, if calculated F-value is more than critical F-value and not rejected, if otherwise. The F-test is carried out at a desired level of significance (LOS) by considering degrees of freedom (DOF) of treatments and error.

CORRELATION ANALYSIS

This is carried out to assess the type, of relation (positive or negative) and depth of relation which range from –1 to +1. This is essential to identify superior variables based on a suitable test of correlations of variables. The analysis is a pre-requisite for screening variables, statistical modeling, refining the predictability of different growth variables and different multivariate problems like multi-collinearity, errors-in-variables, serial dependence of residuals, outliers in data of input and output variables.

Measures of correlation

1. Simple Correlation. It measures the relation between any two variables like tree height, girth, canopy spread etc. It ranges from –1 to +1.

$$r = \frac{\Sigma XY - (\Sigma X)(\Sigma Y)/n}{\sqrt{(\Sigma Y^2 - (\Sigma Y)^2/n)(\Sigma X^2 - (\Sigma X)^2/n)}}$$

2. Partial Correlation (first order). It measures the partial relationship between two variables 'Y' and 'X_1' keeping the effect of a third variable 'X_2' as constant. It ranges between –1 and +1.

$$r_{Y,X_1.X_2} = \frac{r_{YX_1} - r_{YX_2} + r_{X_1X_2}}{\sqrt{(1 - r_{YX_2})(1 - r_{X_1X_2})}}$$

Partial correlation of first order measures the relation between tree height and girth keeping the canopy spread as constant, etc. for all combinations of variables.

3. Partial Correlation (second order). It measures partial relationship between two variables 'Y' and 'X_1' keeping the effects of two other variables 'X_2' and X_3' as constant. It ranges between –1 and +1.

$$r_{Y.X_1.X_2.X_3} = \frac{(r_{YX_1.X_3}) - (r_{YX_2.X_3})(r_{X_1.X_2.X_3})}{\sqrt{(r_{YX_2.X_3})(r_{X_1.X_2.X_3})}}$$

Partial correlation of second order measures the relation between tree height and girth keeping the tree age and canopy spread as constant etc.

4. Multiple Correlation. It measures the correlation of a dependent variable 'Y' with a set of 2 or more independent variables 'X' together. It ranges between 0 and 1.

$$R_{Y.X_1.X_2} = \frac{r^2_{YX_1} + r^2_{YX_2} - 2r_{YX_1}r_{YX_2}r_{X_1X_2}}{\sqrt{1 - r^2_{X_1X_2}}}$$

Multiple correlation measures the correlation between a dependent variable like fruit yield with a set of 2 or more independent variables like tree height, tree girth, canopy spread etc.

5. Rank Correlation. It measures correlation between ranks of two variables instead of the actual observations of variables. It ranges between 0 to 1.

$$r = 1 - \frac{6\Sigma d_i^2}{n(n^2-1)}$$

where, d_i is the difference in ranks.

Rank correlation measures the correlation between ranks of tree species based on height, girth and canopy spread and various other parameters.

6. Correlation Ratio. Correlation ratio 'η' is the appropriate measure of curvi-linear relationship when the relationship between two variables is not linear. If relation is linear then $\eta = r$, otherwise $\eta > r$. It ranges between –1 to +1.

$$h^2_{YX} = [\Sigma(T_i^2/n_i) - T^2/N]/N\ \sigma^2_Y$$

where, $T_i/n_i = \Sigma f_{ij}\ y_{ij}/n_i$

$T/N = \Sigma n_i y_i/Sn_i$

Correlation ratio measures curvilinear relationship of variables like relationship between fruit yield and inputs like organic and inorganic fertilizers, irrigation etc.

7. Intra-class correlation. Intra-class correlation means within class-correlation. Here both variables measure same characteristics. It is the correlation within a variable with respect to some common characteristic. For ex. intra-class correlation between yields of plots.

Suppose there are $A_1, A_2, \ldots, A_n$ families with $k_1, k_2, \ldots, k_n$ members, and let x_{ij} (I = 1, 2, …, n; j = 1, 2, …, k_n) denote the measurement on the jth member in the ith family, then intra-class correlation can be given as

$$r = \frac{\Sigma k_i^2 (x-x)^2 - \Sigma\Sigma(x_{ij}-x)^2}{\Sigma\Sigma(k_i-1)(x_{ij}-x)^2}$$

if $k_1 = k$ (*i.e.*, if all families have equal members), then

$$r = [1/(k-1)][k\sigma_m^2/\sigma^2) - 1]$$

where, σ^2 denotes the variance of X and σ_m^2 denotes the variance of means of families. The intra-class correlation ranges between $-[1/(k-1)]$ and 1.

Intra-class correlation measures the correlation of a variable with in a group or a class with respect to same characteristic. The height of a tree in different weeks, the tree girth in different weeks etc.

Simple regression

It measures the functional relationship between a dependent variable '*Y*' and an independent variable '*X*' with estimates of an intercept (α) and a slope (β). The estimates of α and β can be negative, zero or positive. The linear regression is given as

$$Y = \alpha + \beta X$$

$$\beta = \frac{\Sigma x_i y_i}{\Sigma x_i^2} \qquad \alpha = y = \beta x$$

Simple regression of fruit yield through different inputs provided on different time intervals etc.

Multiple regression

If the dependent variable 'Y' is a function of a set of independent variables '*X*', then the estimates of regression coefficients of different variables (β) along with the intercept (α) are estimated using the matrix algebra. The multiple regression of 'Y' through different independent variables can be given as

$$Y = \alpha + \beta_1 X_1 + \beta_2 X_2 + \beta_3 X_3 + \beta_4 X_4 + \ldots\ldots$$

$$\beta = (X'X)^{-1}\ X'Y$$

The regression coefficients can be negative, zero or positive and would measure the rates of change in '*Y*' for an unit change in the independent variables. Multiple Regression of fruit yield through different input variables like chemical fertilizers, FYM and other organics, irrigation etc.

Polynomial regression

If the dependent variable '*Y*' is a function of linear and other higher order effects of an independent variable '*X*', then the polynomial regression is fitted to quantity the effects of a variable and its significance at different orders for prediction of '*Y*'.

The nth order polynomial regression of '*Y*' can be given as

$$Y = \alpha + \beta_1 X_1 + \beta b_2 X_1^2 + \beta_3 X_1^3 + \beta_4 X_1^4 + \ldots + \beta_n X_1^n$$

The polynomial regression coefficients can be negative, zero or positive and would measure the rates of change in '*Y*' for an unit change in the independent variable. Polynomial regression of fruit yield through inputs used etc.

Principal component analysis

If the different variables are correlated significantly, a principal component model can be derived by computing different uncorrelated principal components as a function of different correlated variables. A principal component model with correlated traits is postulated for calibrating a new set of uncorrelated principals components (Jolliffee, 1986; Bapat, 1993; Dunteman, 1994). Usually, the first two principal components would explain maximum variability in the data. The components are linear and derived in decreasing order of importance. The variables which have got high loadings on principal components could be identified and inferred as dominant variables. Since the variables are related and measured on scales with wide range, the correlations are used for determining principal components. Loadings of traits in a model range between –1 to +1. A trait is considered dominant if its loading is $\geq | 0.7 |$ on a principal component.

TESTS OF SIGNIFICANCE

Different tests of significance are used for testing the differences between sample means and variances of variables observed on different days of sampling, correlation between variables, contribution of different predictor variables (or independent variables) to prediction variables (or dependent variables) as measured by regression coefficients of different horticultural variables. Based on the test procedure, the hypothesis of no differences (null hypothesis) would be rejected if calculated value is more than critical value at a desired level of significance, and not rejected, if otherwise.

1. Testing significance in data of large samples. If X is distributed as Normal with mean μ and variance σ, then $Z = (X - \mu)/\sigma$ is distributed as Normal with mean 0 and variance 1. If $| Z | > 1.96$, then the sample mean is inferred to be significantly different from population mean at 5% LOS. If $| Z | > 2.58$, then the sample mean is significantly different from population mean at 1% LOS.

2. Testing single proportion. If P is proportion of success and Q = 1 – P is proportion of failure. If mean = nP and variance = nPQ, then $Z = [X = nP]/(nPQ)$ is distributed as Normal with mean 0 and variance 1. Same conclusions as above.

3. Testing difference of proportions. Let $p_1 = X_1/n_2$. Mean $(p_1) = P_1$ and Mean $(p_2) = P_2$. Variance $(p_1) = P_1\ Q_1/n_1$ and $V\ (p_2) = P_2\ Q_2/n_2$

$$Z = \frac{(p_1 - p_2)}{PQ[(1/n_1) + (1/n_2)}$$

Z is distributed as Normal with mean 0 and variance 1. Same conclusions as above.

4. Testing a mean in small samples. If X is distributed as Normal with mean μ and variance σ, then $Z = [(x - \mu)/(\sigma/\sqrt{n})]$ is distributed as Normal with mean 0 and variance 1. Same conclusions as above.

5. Testing differences of two means. If X_1 and X_2 are means and $\sigma_1{}^2$ and $\sigma_2{}^2$ are variances based on two samples with n_1 and n_2 observations, then

$$Z = \frac{X_1 - X_2}{\sqrt{(\sigma_1^2 / n_1) + (\sigma_2^2 / n_2)}}$$

is distributed as Normal with mean 0 and variance 1. Same conclusions as above.

6. Testing difference of standard deviations. If σ_1 and σ_2 are standard deviations of two samples with n_1 and n_2 observations from Normal distribution with variances σ_1^2 and σ_2^2, then

$$Z = \frac{s_1 - s_2}{\sqrt{(\sigma_1^2/2n_1) + (\sigma_2^2/2n_2)}}$$

is distributed as Normal with mean 0 and variance 1. Same conclusions as above.

7. Testing the difference between sample correlation (r) and population correlation (p) by making Z-transformation can be given as

$$Z = \frac{Z_1 - Z_2}{\sqrt{n-3}}$$

$$Z_1 = \log_e \sqrt{[(1 + r)/(1 - r)]}$$
$$Z_2 = \log_e \sqrt{[(1+ p)/(1 - p)]}$$

8. Testing the difference between two correlation coefficients 'r_1' and 'r_2' by making Z-transformation can be given as

$$Z = \frac{Z_1 - Z_2}{\sqrt{(1/(n_1 - 3)) + (1/(n_1 - 3))}}$$

$$Z_1 = \log_e \sqrt{[(1 + r_1)/(1 - r_1)]}$$

$$Z_2 = \log_e \sqrt{[(1 + r_2)/(1 - r_2)]}$$

9. Testing the observed correlation '*r*' between two variables against 'zero' can be given as

$$t = \frac{r}{\sqrt{(1 - r^2)/(n-2)}}$$

10. Testing the partial correlation '*r*' between two variables keeping the effects of a third variable as constant can be given as

$$Z = \frac{\log_e \sqrt{(1+r)/(1-r)}}{[1/\sqrt{n-k-3}}$$

11. Testing the regression coefficient (slope) 'b' of an independent variable 'X' can be given as

$$t = \frac{\beta}{\sqrt{s_{y.x}^2 / x_I^2}}$$

12. Testing the regression coefficient (intercept) 'a' can be given as

$$t = \frac{\alpha - \alpha^0}{\sqrt{s_{y.x}^2[(1/n) + (X^2/X_i^2)}}$$

13. Testing the homogeneity of regression coefficients (slopes) of 'k' sets of data (or over different seasons) can be given as

$$H_0 : \beta_1 = \beta_2 = \beta_3 = \ldots = \beta_k$$

$$F = \frac{(B-G)(\Sigma n_i - 2k)}{(B)(k-1)}$$

This is distributed as F with $(k - 1, \Sigma n_i - 2k)$ degrees of freedom

$$G = D - E^2/C$$

B = sum of 'residual sum of squares' of k sets

C = sum of 'corrected sum of squares' of X of k sets

D = sum of 'corrected sum of squares' of Y of k sets

E = sum of 'corrected sum of squares' of products of X and Y of k sets

14. Testing the predictability value (R^2) of a regression model with 'k' independent regressor variables can be given as

$$R^2 = \frac{\beta_1 \Sigma X_1 Y + \beta_2 \Sigma X_2 Y + \ldots + \beta_k \Sigma X_k Y}{\Sigma Y^2}$$

$$F = \frac{R^2}{(1-R^2)} \times \frac{N-K-1}{K}$$

where N is number of observations and K is number of variables

15. Testing the R^2 adequacy of a model with 'p' regressors compared to a model with 'q' regressors where $p < q$

$$\frac{(R^2 q - R^2 p)}{(1-R^2 q)-(n-q-1)} < qF$$

where, F is with $(q, n - q - 1)$ degrees of freedom

$$R^2 p > R^2 a$$

where, $R^2 a = 1 - (1 - R^2 q)(1 + d)$

$$d = \frac{qF}{(n-q-1)}$$

16. Testing the sufficiency of a model with Residual Mean Square Ratio (RMSR) criterion : This is used for testing the sufficiency of a regression model with 'p' regressors compared to a model with 'q' regressors based on F-test as

$$\text{F} = \frac{RSS(p) - RSS(q)}{(q-p)\, RMSS(q)}$$

where, RSS is residual sum of squares; RMSS is residual mean sum of squares. This is distributed as F with $(q - p, n - q - 1)$ degrees of freedom.

17. Percentage relative Efficiency (PRE) of a regression model 'A' compared to a regression model 'B' can be given as

$$PRE\ (A) = \frac{\sigma_B^2 (n+q+1)/n}{\sigma_A^2 (n+q+1)/n} \times 100$$

where, σ^2_A and σ^2_B are Residual mean sum of squares of models '*A*' and '*B*' models

Problems of regression

Multicollinearity. High and significant correlations between different variables compared to over all multiple correlation and predictability of a model. This will result in linear dependence of a variable on another variable and insensitive regression coefficients for prediction. The multicollinearity can be assessed by computing an estimate of c^2 and can be given as

$$\chi^2 = - n - 1 - (1/6)\ (2k + 5)\ \log_e\ (D)$$

where, D = value of standardized determinant

n = number of observations

k = number of independent variables

Analysis of Residuals. This is done for identifying 'Outliers, high leverage and Influential' observations, testing and deletion, and improving the predictability of models.

The residuals can be examined in different forms as normalized, standardized, internally studentised and externally studentised residuals.

Normalised : $a_i = f\ (e_i\ ,\ e'e) = e_i/\sqrt{e'e}$

Standardised : $b_i = f\ (e_i,\ \sigma) = e_i/\sigma$

where, $\sigma = \sqrt{e'e/(n - k)}$

Internally studentised : $r_i = f\ (e_i,\ \sigma\ \text{Ö}(1 - p_{ii}) = e_i/\sigma\ \sqrt{(1 - p_{ii})}$

Externally studentised : $r_i = f\ (e_i,\ \sigma_i\ \text{Ö}(1 - p_{ii}) = e_i/\sigma_i\ \sqrt{(1 - p_{ii})}$

where, $\sigma_i^2 = \sigma^2_{(i)}\ (1 - p_{ii})$

$$\sigma_i^2 = \frac{Y_{(i)}(I - P_{(ii)})Y_{(i)}}{n-k-1}$$

For example, if fruit yield has a predictability (R^2) of only 60%, this means 40% is unpredictable. The unexplained variation as measured by the residuals (difference between observed and expected values of fruit yield of a specie) could be minimized by proper identification and deletion of extreme observations in the data of fruit yield (*Y*) and tree height, girth, canopy spread, and other predictor variables (*X*) and recalibrating the model.

The procedure could be iteratively carried out till there is no significant increase in the predictability estimates of fruit yield or any other dependent variable.

Effects of variables on a model. Examine the effects of independent variables on the dependent variable for their sensitivity and significances, linear dependence and multicollineaity, lack of homogeneity of variances, randomness of independent variables, autocorrelation of variables, extreme nature of variable in its relationships with other variables, violation of the assumption of normality and other aspects.

Normal Distribution : This can be given as

$$f(x) = [1/\sqrt{2\pi}\,\sigma]\; e^{-[(x-\mu)2/2\sigma 2]}$$

(*a*) The curve is bell shaped and symmetrical about the line $x = \mu$

(*b*) Mean, Median and Mode would coincide

(*c*) As x increases numerically, $f(x)$ decreases rapidly, the maximum probability occurring at the point $x = \mu$

(*d*) $\beta_1 = 0$ (Skewness) and $\beta_2 = 3$ (Kurtosis)

(*e*) Linear combination of independent normal variables is also a normal variable

(*f*) Area property : $P(\mu - \sigma < X < (\mu + \sigma) = 0.6826$

$P(\mu - 2\sigma < X < (\mu + 2\sigma) = 0.9544$

$P(\mu - 3\sigma < X < (\mu + 3\sigma) = 0.9973$

(*g*) X-axis is an asymptote to the normal curve

(*h*) The points of inflexion of the curve are given by $x = f(X) = [1/2]\; e^{-[1/2]}$

(*i*) Mean deviation about mean is $\sqrt{2/\pi}\,\sigma = (4/5)\,\sigma$

(*j*) Quartile deviation $= (Q_3 - Q_1)/2 = (2/3)\,\sigma$

(*k*) $\mu_{2r+1} = 0 \qquad (r = 0, 1, 2, \ldots.)$

$\sigma_{2r} = 1.\ 3.\ 5 \ldots (2_r - 1)\,\sigma^2\, r \qquad (r = 0, 1, 2, \ldots.)$

This implies $\mu_1 = 0\;\; \mu_3 = 0\;\; \mu_5 = 0$ .... (odd moments)

$\mu_2 = 1\;\; \mu_4 = 3$ .... (even moments)

Analysis of variance

The differences in effects of input treatments in influencing the fruit yield and soil nutrients could be tested based on the standard analysis of variance (ANOVA) procedure. The treatments which had a significantly higher effect on the soil nutrients and fruit yield compared to others could be identified based on the Least Significant Difference (LSD) criteria (Gomez and Gomez, 1984). Before calibrating a regression model to predict the fruit yield attained by different input treatments, the relationships of the variables could be assessed. The effects of monthly rainfall and soil fertility of nutrients on fruit yield could be assessed based on (*i*) multiple linear regression with original variables and (*ii*) multiple linear regression with principal components derived from the original variables and compare the two approaches for the sustainability of treatments in the long-term study.

REGRESSION ANALYSIS

A regression model of yield attained by an input treatment could be calibrated through rainfall, soil organic carbon (SOC), soil P (SP) and soil K (SK) nutrients observed in different years (Maruthi Sankar, 1986; Draper and Smith, 1998). The regression model of fruit yield could be postulated as :

$$Y = \pm \alpha \pm \beta_1 \text{ (Jul)} \pm \beta_2 \text{ (Aug)} \pm \beta_3 \text{ (Sep)} \pm \beta_4 \text{ (Oct)} \pm \beta_5 \text{ (Nov)} \pm \beta_6 \text{ (SOC)} \pm \beta_7 \text{ (SP)} \pm \beta_8 \text{ (SK)} \quad ...(1)$$

In model (1), α is the intercept and β's are the regression coefficients measuring the change in yield for unit change in a variable. The usefulness of a regression model for yield prediction could be assessed based on the coefficient of determination (R^2) and unexplained variation measured by the prediction error (PE).

PRINCIPAL COMPONENT MODEL

Based on the principal component (PC) analysis as described by Dunteman (1989), Jolliffe (1986) and Krzanowski (1979), the PCs of each treatment could be determined through monthly rainfall received from July to November; SOC, SP and SK nutrients. Based on the PC model, the ith PC 'P_i' could be postulated as

$$P_i = \pm a_{i_1}\text{(Jul)} \pm a_{i_2} \text{ (Aug)} \pm a_{i_3} \text{ (Sep)} \pm a_{i_4} \text{ (Oct)} \pm a_{i_5} \text{ (Nov)} \pm a_{i_6} \text{ (SOC)} \pm a_{i_7} \text{ (SP)} \pm a_{i_8} \text{ (SK)} \quad ...(2)$$

In (2), a_{ij} are loadings of variables on 'P_i', which are chosen such that they satisfy two conditions, *viz.*, (*i*) the PCs are orthogonal with respect to each other; and (*ii*) 1st PC, say P_1 will account for maximum variance of the variables; 2nd PC will account for maximum of the remaining variance in variables after eliminating the variance accounted by 1st PC; and etc. An eigen value 'l_k' indicates the variance extracted by the kth PC and can be expressed as percentage of total variance of all components, *i.e.* $P_k = (l_k/k) \times 100$. The computational procedure of 1st PC is repeated for the 2nd PC and all subsequent PCs. The 1st PC would always have a higher eigen value than the 2nd PC; the 2nd PC has a higher eigen value than the 3rd PC and so on.

Based on the analysis, we determined (*i*) the significant number of PCs with eigen value of more than 'one' to extract maximum variance in the data; (*ii*) variables which have a significantly higher loading of more than + 0.70 or less than – 0.70 on PCs; and (*iii*) per cent of variance explained by a PC. We can identify a superior treatment for which the PC model (*i*) explained maximum variance in the data; (*ii*) provided maximum number of variables with a significantly higher loading on PCs.

Regression analysis of yield of treatments through principal components

Let P_1, P_2, P_3, ..., P_k are the principal components (PCs) derived from the 8 independent variables as described in (2). The yield of a given treatment can be fitted using the PCs fulfilling the above specified criteria as under.

$$Y = \alpha + \sum_{i=1}^{k} \beta_i P_i D_i + \varepsilon$$

where, α is the intercept

P_i is i^{th} the PC

b_i is the regression coefficient corresponding to i^{th} PC

D_i is a dummy variable which takes a value of '1' if the ith PC is included in the model and '0' otherwise. The model could be evaluated based on the coefficient of determination (R^2) and prediction error over a time period.

Sustainability yield index

The sustainability yield index (SYI) of an input treatment could be measured as a ratio of the "difference between mean yield and prediction error" and "maximum yield" attained by the treatment in any year in the study period. The SYI of treatment 'i' could be derived by using the mean yield (A_i) of a treatment over years; prediction error (F_i) derived based on the model of ith treatment; and maximum yield (Y_{max}) attained by any treatment in the long-term study. The SYI 'η' of treatment 'i' is derived as

$$\eta_i = [(\tilde{A}_i - \Phi_i)/(Y_{max})] * 100 \qquad ...(3)$$

In (3), the mean yield of a treatment would get detrended by elimination of the unexplained variation as measured by the regressor variables in the multiple regression model and principal components in the PC model through rainfall, SOC, P and K nutrients over years. The detrended yields are compared with the maximum yield attained by a treatment in any year in the study period for assessing the sustainability of the respective treatment.

REFERENCES

Draper, N.R. and Smith, H. (1998). Applied Regression Analysis. Wiley Publishers Inc., New York.

Gomez, K.A. and A.A. Gomez. (1984). Statistical procedures for agricultural research. Wiley Publications Inc., New York.

Jolliffe, I.T. (1986). Principal component analysis. Springer – Velag, New York.

Krzanowski, W.J. (1979). Some exact percentage points of a statistic useful in analysis of variance and principal component analysis. Technometrics. **21** : 261-263.

Maruthi Sankar, G.R. (1986). On screening of regression models for selection of optimal variable subsets. *Journal of Indian Society of Agricultural Statistics,* **38**(2) : 161-168.

●●●

5

Elevating Economic Returns in Fruit Production through Low Input and High Return Strategy Introduction

N.N. REDDY AND ABHIJIT DEBNATH

Excessive use of chemical fertilizers and pesticides as a mean of intensive cultivation to boost up our food production have caused considerable damages to our soil health and the environment. Over the last half century, conventional approaches to crop production have relied heavily on manufactured fertilizers and pesticides to increase yields, but they have also degraded water quality and posed threats to human health and wildlife (Dubrovsky *et al.*, 2010). Consequently, attention is now being directed toward the development of crop production systems with improved resource use efficiencies and more benign effects on the environment. Balancing productivity, profitability, and environmental health is a key challenge for agricultural sustainability (Foley *et al.*, 2011). Orchard cultivation refers to the careful management of the orchard soil in such a way that the soil is maintained in a good condition suitable to the needs of the tree with least expenses. This involves maintenance of the physical condition of the soil, its moisture and nutrient content. A good system of orchard cultivation should ensure :

1. Very little disturbance to soil for preventing soil erosion.
2. Weed control for saving in moisture and nutrients.
3. Reduced cost of cultivation utilizing low inputs and high return strategies.

The continuous supply of organic material feeds a huge number of soil organisms and provides an ideal environment for them. As a result the soil becomes soft and capable of taking up and storing large quantities of nutrients and water. Organic farmers give importance to the improvement of soil fertility.

MEANS OF LOW-INPUT PRODUCTION

Low-input farming can be defined as systems managed with reduced use of inputs, usually resulting in a system that suffers from some type of limitation or stress, commonly nitrogen and phosphorus deficiencies or inadequate water supply, that ultimately cause yield losses. These systems are not necessarily organic in practice (as defined by the USDA), since conventional low-input and organic high-input operations are not unheard of around the world. However, low-input is often associated and used as a synonym for organic production systems, especially in developed countries. Low-input systems have reduced use, but not elimination of fertilizers (either from inorganic or organic sources), or pesticides and herbicides (biological, inorganic or organic).

Low input farming systems. "Seek to optimize the management and use of internal production inputs (*i.e.,* on-farm resources) and to minimize the use of production inputs (*i.e.,* off-farm resources), such as purchased fertilizers and pesticides, wherever and whenever feasible and practicable, to lower production costs, to avoid pollution of surface and groundwater, to reduce pesticide residues in food, to reduce a farmer's overall risk, and to increase both short and long-term farm profitability" (Parr *et al.,* 1990).

In fact, the term low-input referred to purchasing few off-farm inputs (usually fertilizers and pesticides), while increasing on-farm inputs (*i.e.,* manures, cover crops, and especially management). Thus, a more accurate term would be different input or low external input rather than low-input" (Norman *et al.,* 1997).

LOW-INPUT SUSTAINABLE AGRICULTURE

Inputs in crop cultivation include such factors as labour, tools, material inputs, farming techniques, information *etc.* Most fertiliser was of the traditional *(i.e.,* organic) type : manure, crop residues, urban waste and (in two cases) ash.

Type of input : No inputs, chemical fertilisers, manure as fertiliser, crop residue as fertiliser, urban waste as fertiliser, ash as fertiliser, chemical pesticides, chemical insecticides, irrigation, improved seeds/seedlings.

Inputs can be classified in various ways. An initial division is between traditional and modern inputs. Manure, crop residues and local seeds can be considered as traditional inputs, while the three chemicals plus improved seeds can be seen as modern inputs. The question behind this classification is whether the use of modern inputs leads to higher yields. A second classification is between chemical inputs and non-chemical inputs, which is important for environmental reasons. In this context, one can also distinguish so-called sustainable (or environmentally friendly) inputs that are organic and can be recycled and may lead to reasonable yields: manure, crop residues and improved seeds. Finally, irrigation is an input that stands on its own: without sufficient water no harvest is possible. The relationship between these categories of inputs on crop yields is dealt with as detailed below :

Another way of input classification : traditional inputs, modern inputs, chemical inputs, sustainable inputs, irrigation etc.

It is hypothesised that the use of inputs varies with certain plot characteristics (size, location, distance, ownership), household characteristics (income, size) and characteristics of the person responsible for crop cultivation (sex, marital status, educational level, age). To simplify matters, three mutually exclusive categories of material inputs are compared : chemical inputs, sustainable inputs and irrigation, after which 'hired labour' and 'assistance' are considered. Many factors interact to determine crop production. Cropping systems have evolved or been developed to achieve high yields, relying on practices that eliminate or minimize yield reducing factors. However, this is not entirely the case in many developing countries where subsistence farming is common. The soils in these countries are mainly coarse-textured, have low water holding capacity, and are low in fertility or fertility declines rapidly with time.

LOW INPUT APPLE PRODUCTION

Weibel *et al.,* 2006 reported improved organic and low-input apple production using rootstocks tolerant to weed competition. In low-input and organic, sustainable apple (*Malus domestica*) production, tree development during the first three years after planting is often inadequate. Using traditional vigorous rootstocks to overcome this problem is not a solution for modern fruit production because yield sets in too late and final tree size is too big to produce economically. In 1998, a trial was initiated at three sites in Switzerland to test 10 rootstock selections grown under 100 per cent organic, extensive management : Czech rootstocks J-TE-G, J-TE-E, J-TE-F and J-OH-A; Budagovsky 9, Cepiland (Pajam® 2), P22, Supporter 2, M-7 (as a reference for an over-vigorous rootstock) and M-9 Fleuren 56 (as a reference for a modern dwarfing rootstock) with three scab resistant cultivars : (*i*) 'Resi'; low vigour, small fruit; (*ii*) 'Resista'; vigorous, poor branching; (*iii*) 'Ariwa'; balanced vigour, well branched. We assessed growth vigour, branch development, yield, fruit size and quality, suckers and burr knots, plant health, and nutrient uptake in both leaves and fruit. There was a highly significant influence of rootstock and variety and regular interactions of rootstock with cultivar. After the 6th growing season, highest total cumulative yields occurred with M-7 (27.8 kg/tree), Supporter 2 (19.8 kg/tree) and Cepiland (15.0 kg/tree) whilst lowest occurred with J-TE-F (6.9 kg/tree). The rootstock-induced differences between lowest and highest yield was 416. With cumulative relative yield (kg per mm stem-circumference) M-7 and Supporter 2 performed best (2.13 and 1.94 kg/mm), followed by Fleuren 56 and Cepiland (both 1.68) and P22 (1.52) with the difference between lowest and highest rootstocks being 231. In 2005, with 153 mm stem circumference M-7 was 29 more vigorous than Supporter 2 (118), Cepiland (105) or Fleuren 56 (83). The tree vigour of Supporter 2 was judged almost ideal getting an average of 99 rating points (where 100 points would be the optimum) as well as in terms of feathering and leaf condition. Relating to nutrient uptake and fruit quality, significant differences were frequently found that correlated with tree performance and fruit quality. Under Swiss conditions, Supporter 2 seems to be a good alternative to M-9 type rootstocks for low-input and organic apple production.

ELEVATING ECONOMIC RETURNS IN FRUIT PRODUCTION INCLUDES

(*i*) Planning of an orchard for high economic return
(*ii*) Fruit variety selection
(*iii*) Use of organic inputs
(*iv*) Implication of conservation horticulture facts in fruit production
(*v*) Soil management and water conservation measures
(*vi*) Application of Cropping systems in orchard for space utilization
(*vii*) Orchard Floor Management
(*viii*) Integrated Nutrient Management
(*ix*) Introduction of High Density Plating and Meadow Orcharding
(*x*) Plant protection measures by means of low inputs
(*xi*) Weed management
(*xii*) Proper intercultural operations for higher yield of fruits
(*xiii*) Proper labour management
(*xiv*) Improvement of farmer's technical know how

I. Planning of an orchard for higher economic returns

A careful plan of the orchard should have the following components :

1. Optimum spacing to accommodate maximum number of trees per unit area.
2. Stores and office building in the orchard should be constructed at the centre.
3. Wells at convenient places in different parts one @ one well for 2 to 4 hectares.
4. Each kind of fruit should be assigned in a separate block.
5. Fruits ripening at the same time should be grouped together.
6. Every third tree in every third row should be a pollinator in deciduous fruit trees.
7. For every 30 m length of irrigation channel, 7.5 cm slope gradient should be given.
8. The clearance between wind break and first row of trees has least space for the road.
9. Short growing trees at the front and tall at the back for easy watch and ward.
10. Evergreen trees should be in the front and deciduous ones behind.
11. Fruits attracting birds and animals should be close to the watchman's shed.

12. Live fencing should be drought resistant, easy to propagate, quick growing, dense foliage, stand severe pruning and thorny. *Agave, Prosopis juliflora, Pithecolobium dulce* and *Thevetia* if closely planted in 3 rows would serve as a good live fencing.
13. A wind break with rows of tall trees has its effect four times as great to its height.

The most effective windbreak is a double row of tall trees alternately placed. Trees suitable for windbreak should be erect, tall and quick growing, hardy and drought resistant and mechanically strong and dense to offer maximum resistance to wind. The trees suitable for growing as wind breaks are *Casuarina equisetifolia, Pterospermum acerifolium, Polyalthia longifolia, Eucalyptus globulus, Grevillea robusta, Azadirachta indica* etc.

While selecting a site for establishing a new orchard, the following need to be considered :

1. Suitability of soil, its fertility, the nature of subsoil and soil depth.
2. Site must have proper drainage and no water stagnation during rainy season.
3. Irrigation water must be of good quality.
4. There must be proper transport facilities either by road or rail within the reach.
5. Whether the climatic conditions are suitable for the fruits spp and whether site is free from cyclones, frost, hailstorms and strong hot winds.
6. Whether there are seasonal gluts or over production in any period of the year.
7. Whether there is assured demand in the market for the fruits to be grown.
8. Whether the orchard is a new venture or whether there are already other growers.
9. Availability of labour (Kumar, 1997).

II. Fruit variety selection

Selection of material for planting is determined by a number of factors. These include :

1. Growing conditions.
2. Pest and disease resistance.
3. Fruit quality parameters (such as size, firmness and keeping ability).
4. Maturity dates.
5. Market requirements.

Select varieties for the available markets, not *vice-versa*. Contact wholesalers and retailers of organic produce to determine the lines in which demand exceeds current supply. Select varieties that are adaptable. Choosing genetically resistant cultivars is a very important control measure for organic farmers especially with regard to disease management. In some cases, such as bacterial spot in peaches, cultivar resistance is the best or only control.

III. Use of organic inputs

Essential features of organic inputs-1. Use of organic manures like FYM, compost, vermicompost and coir compost etc. 2. Use of bio-fertilizers. 3. Use of green manures and grain legumes. 4. Non-chemical weed management. 5. Use of botanicals and bio control agents in the control of crop pests. In India, organic farming in horticultural crops especially in vegetable and spices crops are slowly picking up, wherever premium prices are available for organically grown quality produces. For conversion of a conventional field to organic, first step is to build up the lost fertility of the soil. Biological resources such as biofertilizers and other microbiological inputs have also attracted lot of attention and are being promoted on large scale. Under various other schemes of Central and State Governments assistances are available for setting of vermicompost production facilities.

Bio-fertilizers or microbial inoculants are carrier based ready to use live bacterial or fungal formulations. To ensure a good quality bio-fertilizer, a formulation should possess following traits.

(*a*) The product must be carrier based or liquid formulation, capable of holding very high population of specific micro-organisms for sizeable period of time.

(*b*) In case of carrier based formulations the product should have 30-50 of moisture throughout the shelf life period to sustain microbial population.

(*c*) For carrier based formulations the microbial population should be in the range of 10^7 to 10^9 cells/g of moist product. In case of liquid formulations it should be 1 × 108 to 1 × 1010 during the entire period of shelf life.

(*d*) It should be free from other contaminating micro-organisms.

(*e*) The microbial strain present in the product should be able to produce adequate nodulation in case of *Rhizobium*, be able to fix at least 10-15 mg of N/g of carbon source used in case of free living N_2 fixers and be capable of solubilising significant quantity of fixed soil P.

(*f*) It should have sufficient shelf life (minimum 6 months for carrier based and 12 months for liquid).

It is a production system which avoids or largely excludes the use of synthetically compounded inorganic chemicals. This system entirely relies on crop rotation, crop residues, animal manures, legumes, green manures, off-farm organic wastes, biofertilizers, mechanical cultivation etc. and aspects of biological pest control to maintain soil productivity and tilth to supply nutrients and to control insects, weeds and other pests. This system is often, referred as 'biological farming' 'regenerative farming' and 'sustainable farming' 'eco-friendly farming' etc.

Low cost of production by utilizing organic inputs

Organic agriculture has triggered a controversial debate in the last decade, most importantly because it shed light on the darker sides of chemical-intensive conventional farming by offering an alternative. It is not clear whether organic agriculture could be economically attractive enough to trigger wide spread adoption. If organic farming offered a better environmental quality, and potentially healthier foods, but not sufficient economic returns to the majority of farmers, it would obviously remain a luxury way of food production available to a very tiny fraction of farmers.

An Indo-Swiss research team compared agronomic data of 60 organic and 60 conventional farms over two years (Eyhorn *et al.*, 2005) and came to the conclusion that cotton-based organic farming is more profitable. The updated version of the study surveying 125 organic cotton farmers concluded that 95 per cent of respondents witnessed their agricultural income raises after adopting organic agriculture (MacDonald, 2004).

Reganold *et al.* (1987) compared the long term effects (since 1948) of organic and conventional farming on selected properties of the same soil. The organically farmed soil had significantly higher organic matter content, thicker topsoil depth, higher polysaccharide content, lower modulus of rupture and less soil erosion than the conventionally farmed soil.

Haas *et al.* (2005) conducted a study in Hamburg, Germany to estimate the environmental impact of conversion to organic agriculture by using a Life-Cycle Assessment (LCA) method.

TABLE 5.1
Approximate Nutrient Composition of Various Organic Fertilizers and Soil Amendments

	N	P_2O_5	K_2O
		(dry weight basis)	
Dairy manure	2.1	3.2	3.0
Beef manure	1.2	2.0	2.1
Poultry manure	3.0	5.0	2.0
Composted yard waste	1.3	0.4	0.4
Animal tankage (dry)	7.0	10.2	1.5
Alfalfa hay	2.5	0.5	2.5
Blood meal	13.0	2.0	1.0
Fish meal	10.0	6.0	0
Kelp/seaweed	1.5	1.0	4.9
Soybean meal	7.0	1.2	2.0
Bone meal (raw)	3.0	22.0	0
Bone meal (steamed)	1.0	15.0	0
Cottonseed meal	6.0	3.0	1.5
Wood ashes	0	2.0	6.0
Rock phosphate (total P_2O_5)	0	20-32	0
Colloidal phosphate (total P_2O_5)	0	25	0
Greensand (total P_2O_5 and K_2O)	0	1.3	4-9.5
Granite dust (total P_2O_5 and K_2O)	0	0	3-5
Sul-Po-Mag (langbeinite)	0	0	22
Potassium sulphate	0	0	50

(Rosen and Bierman, 2005)

Energy efficiency by organic fruit production

A study of the sustainability of apple production systems showed that in comparing a conventional farming system to an organic method of farming, the organic system is more energy efficient (Reganold *et al.,* 2011). However, conventional farming had higher total yield Dalgaard *et al.* (2001). Conversely, another study noted that organic wheat and corn production was more energy efficient than conventional methods while organic apple and potato production was less energy efficient than conventional methods Pimental *et al.* (1983).

A study done with apple orchards in the state of Washington found that organic orchards found to be at least 7 more energy efficient. Reganold *et al.*, Glover, J.D., Andrews, P.K., Hinman, H.R. (2001). A sign outside of an organic apple orchard in Washington reminding orchardists not to spray pesticides on these trees.

High water-use-efficiency

The water holding capacity (WHC) of modern/conventional farm ranged from 28-33 whereas, in case of organic farm it ranged from 42-47 (Shiva *et al.*, 2004). In general, indigenous wheat varieties need 12 inches of irrigation where as the hybrid varieties require at least 36 inches of water (Shiva *et al.*, 2004).

IV. Implication of conservation horticulture in fruit production

Conservation horticulture is a system of integrated management of soil, water and biological resources combined with external inputs. It is a basic management tool (Lal, 1989). "To be economically attractive for farmers, Conservation Horticulture must be perceived to provide a net economic benefit either through lower production costs or higher crop yields, higher net returns, lower business risks, or some combination of these." (Sanchez-Giron *et al.*, 2004).

TABLE 5.2

Key agronomic and environmental challenges in the main agro-climatic regions

Agro-climatic	Objectives for Conservation Agriculture conditions
Dry rain-fed	Restrict competition for water by weeds, Increase infiltration and control evaporation, Reduce soil degradation and restore or maintain soil fertility
Moist rain-fed	Avoid water logging
Irrigated	Restrict competition for water from weeds, Increase infiltration and control evaporation, Reduce soil degradation and restore or maintain soil fertility, Reduce off-site movement of agrochemicals and sediments
Hilly	Improve soil quality of reduced rooting zone, Increase infiltration, Avoid mass movement of soils, Reduce off-site movement of agrochemicals and sediment

The water conservation is inseparable from soil conservation. The measures useful in controlling are discussed below.

Contour bunds. Bunds are either mechanical or vegetative barriers created across the slope. The purpose is to divert the excess run-off during rain to the waterways and to retain eroded soil. These bunds on steep slopes are created by way of excavating parabolic channels (0.3 m top and 0.2 m deep) on contours and keeping the dug out soil in form of a bund at the lower edge of the channel. These bunds require care in maintenance during first 2 years. Our experience showed that vegetative barrier alone will not serve the purpose on steep slopes. The vertical interval of these bunds may vary from 0.5 to 5 m depending on the land use, soil depth and slope of the land.

Bench terrace. Bench terraces are flat beds constructed across the hill slope; spaces between two contours are levelled by cut and fill method. In micro-watershed involving steep slopes, experiences showed that only at foot hills few benches may be constructed to produce food crops through intensive cropping. The vertical interval of such terraces should not increase more than one meter. Such measures can be adopted where the soil depth is more than 1.0 m.

In case the entire hill slope is to be converted into benches, the construction should start from foot hill. Half moon terrace: These are level circular beds having 1 to 1.5 m diameter cut into half moon shape on the hill slopes. These beds are used for planting and maintaining sapling of fruits and fodder trees in horticulture/agro-forestry land uses.

V. Soil and water management measures

1. Clean culture. This type of cultivation is extensively followed in India. This involves regular ploughing and removal of weeds. The clean culture has many disadvantages. They are :

(*i*) Humus will be completely depleted rapidly due to frequent cultivation.

(*ii*) Frequent cultivation injure feeder roots, trees may be short lived or stunted in growth.

(*iii*) Clean cultivation aids in more aeration leading to the depletion of nitrogen.

(*iv*) Hard pan is created in the soil.

(*v*) Frequent cultivation causes more soil erosion.

The above can be minimized by avoiding deep, frequent cultivation when the soil is too wet.

2. Clean culture with cover crops. It is the best to plant a green manure crop between the trees early in the rains and plough it into the soil towards the end of monsoon season. In India, green manure crops like Sun hemp, Cowpea, Daincha, Lupins etc. are more commonly used. Legume cover cropping in grape, mango, guava and other fruit crops is becoming a common practice in the management of orchards. Cowpea and French beans grow well under guava and sapota tree. In some places to prevent soil erosion, certain permanent cover crops like *Calapogonium muconoides, Centrosema pubescens* and *Peuraria phaseoloides* are raised in the alley spaces. They are leguminous crops, establish in a short period and dry up during summer to conserve moisture. With summer showers they come up again because of their profuse seeding habit and spread themselves as a vegetative mat by the time the heavy monsoon starts pouring in. Such permanent cover cropping is a common feature in rubber plantations of Kerala and Kanyakumari district.

Mulching. This is one of the important soil management practices adopted in certain countries. Crop residues like straw, cotton stalks, leaves, saw dust, pine needles, coir dust arid other materials like polythene films or certain special kinds of paper are spread in the tree basins and in inter spaces between trees. Main objective of mulching is to conserve soil moisture and to control the weed growth.

The other **advantages** of mulching are :

1. Keeps soil cool in day; warm at night hours
2. Reduces surface run-off
3. Adds humus to the soil
4. Prevents soil erosion
5. Fruits are protected and kept clean since they fall on the mulches
6. It allows the absorption of more rain water and
7. It reduces irrigation frequency.

The following are some of the **disadvantages** :

1. Dry materials used as mulches encourage the risk of fire and consequent damage to trees.
2. Thick mulches may act as places for mice and rodents to live and multiply. They may cause damage to tree trunks and roots by eating the bark and burrowing to the land. The mulching materials should be placed too close to the tree trunk and it should be spread in such a way that they give a good cover to the root system of the trees.

Fig. 5.1. Mulching in Litchi tree basin

4. Sod. In this method, permanent cover of grass is raised in the orchard and no tillage is given. This type of orchard cultivation is followed in USA and Europe. This may be useful in slopy lands for preventing soil erosion. But they compete for soil moisture and available nitrogen. The drawbacks of this system are the need for increased manuring and water application. They are harmful to shallow rooted trees. Hence sod may be useful with deep rooted trees because soil moisture will be very low on the top layers.

5. Sod mulch. This is similar to sod with the only difference is that the vegetation is cut frequently and the cut material is allowed to remain on the ground. This is slightly better than the previous one, as the moisture loss is not so great as in sod. In both sod and sod mulch, more nitrogen should be applied to the fruit trees than usual application because the vegetation utilises more soil nitrogen.

Grassed waterways. These are channels laid out preferably on natural drainage lines in the watershed. As far as possible, natural courses should be used without much disturbances for draining out the excess water. At appropriate locations, stilling basin (water pools) should be created with land use of earthen and boulder pitched bunds for temporary detention of run-off water. This structure will also serve the purpose of energy dissipation of flowing run-off water.

Water harvesting ponds. Dug out cum embankment type of water harvesting structure can be used for creating seasonal and perennial ponds at the foot of the micro-watershed for irrigation. On hill slopes, soil and water conservation measures may be followed with horticulture land use system. Other steps required are : Contour bund at 1.0 m to 5 m. vertical interval in all the land uses with common grassed waterways. Bench terraces towards the foot hills for growing vegetable crops (Vertical interval should not increase over 1.0 m). Half-moons terrace (1 to 1.5 m dia.) for fruit crops. Water storage at appropriate location for irrigation/fish purposes.

Watershed based land use system. Watershed management is the development and management of the watershed resources in such a manner so as to achieve optimum production, which can be sustained without causing deterioration in the resources base or disturbing the ecological balance.

Principles of watershed management. Our experiences at CRIDA proved that with the watershed based concept, it is possible to manage the hill slopes effectively while utilizing local natural resources. Land has to be developed to receive the rainwater in such a way so as to utilize the maximum for plant growth and lead the rest into storage either surface water storage or ground water storage for later use by man and animals.

VI. Application of different cropping systems in an orchard

Intercropping. In the case of long duration horticultural crops like tapioca, turmeric, ginger and banana some area between adjacent plants wiii be remaining unoccupied by the main crop for few months. It naturally appeals to the grower to get some return from this vacant land especially when he is getting no return in the early periods. The practice of growing any economic crop in alley spaces of the fruit trees in the first few years or in the unoccupied spaces of the long duration crop in the early periods is referred as intercropping. The following important principles should be observed while growing intercrops.

- Intercrops should not occupy the rooted area of the fruit trees.
- Soil fertility should be maintained or improved when intercrops are grown.
- Water requirements of the intercrops should not clash with the main fruit trees.
- Intercrops should be selected with reference to their effect on soil moisture. Vegetables are the best inter crops when compared to millets. Many growers prefer some quick growing fruit trees to grow as intercrops. The fillers should be removed after a few years usually immediately after the main fruit trees have commenced bearing.

Recommended intercrops for important horticultural crops in the orchards

Crop	Age	Intercrop
Mango	Up to 7 years	Leguminous vegetables, Papaya (filler)
Grapes	Up to 8 months	Snake gourd or bitter gourd in pandal
Apple, pears	Up to 5 years	Potato, Cabbage
Banana	Up to 4 months	Sun hemp, onion
Tapioca	Up to 3 months	Onion, beans, lab-lab, black gram
Turmeric	Up to 3 months	Small onion, coriander
Arecanut	Up to 10 years	Pineapple
Coconut	Up to 3 years	Banana, tapioca, vegetables

Mixed cropping. It refers to the practice of growing certain perennial crops in the alley spaces of the main perennial crops. Extensive research conducted by CPCRI, Kassargode on mixed cropping in coconut and arecanut plantations showed that cocoa, pepper, cinnamon, clove and nutmeg can be grown as mixed crops in coconuts while nutmeg and clove as mixed crops in between four arecanut palms on alternate rows. In all the above cases, increase in yield (up to 10) is obtained in the main crop due to the synergistic effect of the crop combinations arising out of beneficial micro organisms in the rhizosphere and the more availability of major nutrients in the active root zone of the crop mix as compared to the pure stand.

Multitier system of cropping. Certain horticultural plants like coconut and arecanut are grown for about 50 years in a particular land. It takes nearly 4 to 7 years for the above trees to reach the bearing stage. Adequate alley spaces (nearly 75) are available in between these trees and being the palm trees, their root system will not also spread beyond one metre in diameter.

Intercropping and mixed cropping involve jointly multitier system of cropping and is defined as a compatible companion of crops having varying morphological frames and rooting habits, grown together in such a manner that their canopies intetercept solar energy at varying heights and their roots forage the soil at different zones. The main principle here is that the land, water and sunlight should be effectively used. An ideal combination of crops for multitier cropping in coconut and arecanut plantations is as follows.

Tier	Crop
First (Top)	Coconut or arecanut
Second	Pepper trained over the trunk of coconut or arecanut trees
Third	Cocoa or cloves planted at the centre of four arecanut or coconut
Fourth (ground)	Pineapple, ginger and dwarf coffee

(N. Kumar, 1997)

Filler Fruit Crop based Multi-tier System. To maximize the per unit area output from litchi orchards growers prefer some quick growing and early bearing fruit trees. In some areas in a properly spaced litchi orchard there is scope for guava trees which begin to bear in second year and produce a good yield up to 8-10 years other smaller trees, may be papaya, citrus or phalsa which have been found suitable as filler plant in litchi (Fig. 5.2). In no case the roots of the filler trees occupy the soil where the roots of the permanent trees are concentrated.

Fig. 5.2. Litchi + Guava + Cowpea system in NEH region

Horticultural land use (with fruit crops only). The land use under pure horticultural orcharding system has high potentiality in the region where more area (jhum land) is available at a place. This system of cultivating the land will be highly profitable in the long run and area will be developed as fruit growing belt. The water and soil management practices are to be followed in a systematic manner so that the soil loss can be checked to a considerable stage (Yadav *et al.,* 1999).

VII. Orchard Floor Management

In terms of controlling erosion, a system that results in full or close-to-full ground coverage is the best. Where adapted, orchard grass and other cool season grasses are often recommended because they go dormant during the heat of the summer, thereby minimizing competition with the fruit crop for water. Mustards, buckwheat, dwarf sorghum and various members of the Umbelliferae and Compositae families support substantial numbers of beneficial insects without attracting as many pests.

After a planting is established, mulching with organic materials such as straw, leaves or sawdust can provide significant weed suppression. Generally, mulch should be kept well away from the trunk to reduce damage from voles. This is especially important in winter.

Cover crops. Inter-row cover crops can provide additional nutrients to the crop, increase soil organic matter and improve orchard biodiversity. Several techniques can be adopted.

Green manure crops. Green manure crops are grown specifically to be cultivated back into the soil to build up soil organic matter and nutrients. They are especially useful during orchard establishment.

Permanent sods. This favours the development of plant roots and soil biology; and helps to retain good soil structure. Ideally, a sod consists of a range of perennial plant species. For example it may contain a mixture of :

- Grasses (for example ryegrass or fescue), efficient in obtaining potassium from the soil and able to utilise excess organic nitrogen;
- Legumes (for example clover or lucerne), able to add 40-140 kg per hectare per year of nitrogen to the soil reservoir and also 'feed' the grasses and the crop;
- Herbs (for example comfrey and chicory), high in essential elements with deep roots capable of bringing up leached elements those are otherwise unavailable to the crop.

Semi-permanent sods. These consist of annuals which die down or are slashed in early summer to form mulch in the inter-row area. Suitable species include annual ryegrass, oats, lupins, vetch and field peas. Alternatively, warm-weather crops may be planted late. These die off in cold weather, allowing a mulch to form, and result in more soil moisture in spring than if a winter-hardy cover crop had been planted. There is less competition from semi-permanent sods for water and nutrients, and they require less frequent mowing, but they also offer less diverse orchard fauna and flora (Robyn, 2008).

Jennifer *et al.* (2009) worked on developing Organic and reduced-input tree-fruit production options for the Wasatch Front, Utah Germany and found beneficial effects of trees, living and non-living mulches and nutrient sources which alter pest dynamics, water use, soil quality, fruit quality and economics.

Fig. 5.3. Alyssum living mulch did not persist beyond the second growing season.

Fig. 5.4. Organic treatment with tillage weed control. Tree growth equalled the conventional check.

Fig. 5.4. Organic treatment with weed fabric mulch as weed control. Tree growth equalled the conventional check.

Benefits of Proper Orchard Floor Management

1. Due to wide spacing and developing root patterns, the large unutilized interspaces can be exploited for growing inter and mixed crops successfully.
2. The soil fertility can be enhanced by careful selection of intercrops.
3. Cover crops save the orchard soil from being eroded during the rainy season. When the cover crops are turned into the soil, the organic matter of the soil is increased.
4. Water-holding capacity of the soil is increased and the biological complex of the soil is also improved.
5. Cover crops prevent weed growth; protect the young plants from wind and even from cold injury.
6. Cover crops also a source of mulching after they have been harvested.
7. Constant use of intercrop may reduce the soils' phosphorus and potassium content while increasing their nitrogen content. Because of this, small quantities of P and K should be incorporated to offset these losses.

Selection of Crop. Where the investment is large and expenses are high, intercrop would do much help or produce profit during first few years. Factors Determining Choice of Intercrops—

1. Orchards nearer to cities or towns, better to grow vegetables. For distance orchards crops like onion and potato can be grown during young stage.
2. Select the crops which are well adapted to the climatic and soil conditions of the area. Legumes are better choice for soils poor in nitrogen.
3. Water requirements of the intercrops should as far as possible coincides with the requirement of litchi plants. Such intercrops should be selected that do not exhaust the nutrients and moisture from the soil which are essential for the growth of litchi plants.
4. For young orchards, vegetables which require abundant sun-light can be selected. For old orchards shade loving crops are preferred. It may be kept in mind that the intercrop should not affect the management of the main crop or interfere with the cultural practices. It is not desirable ordinarily to grow grain or other exhaustive crops like sugarcane, because of the exhaustion of nutrients and moisture.
5. Crops like wheat, oat, grain, oilseeds and radish (for seed) may not be grown as intercrops in grownup litchi orchard. This is mainly because the period of seed production and subsequent development of seed in the intercrop generally coincides with the period of active fruit development and during the period the intercrop extracts maximum amount of moisture and nutrients from the soil.
6. Perennial or exhaustive crops should be discouraged as an intercrop in the orchard. This may have devitalizing effect on the growing trees. For example, sugarcane, pigeon pea, maize, jowar should invariably be excluded from an intercropping programme in the litchi orchard.

7. The crops that may grow tall and have a tendency towards excessive growth should be discouraged. Trailing vegetables or vine crops are not desired as intercrops. Growth habit must be such that it does not affect the growth of litchi crop directly.
8. Certain pest and disease problems are enhanced by growing two or more crops together. It is also noted that intercrops favour some pests and pathogens since they serve as alternative hosts for them. Such crops may not be allowed to grow together.
9. Vegetables that have their roots within 25 cm depth of the soil are considered good for intercropping. (Tomato, onion, cauliflower, beans, radish, palak, etc.)
10. It is preferable to take leguminous crops for cover cropping so that nitrogen fixation in the orchard soil is facilitated. Another advantage of cover crops is that they will prevent the growth of weeds. During rainy season (kharif) greengram, blackgram, cowpea and guar are preferable. During winter season (rabi) it is desirable to take chick pea, pea, lentil, mustard, tomato, potato, taba bean, etc.
11. Some crops like sun hemp in light soils and Dhaincha in heavy soils may also be grown to protect the orchard soil from erosion and also for enriching the soil fertility. These crops are sown in beginning before rainy season and ploughed back after 45-50 days before or just before flowering.
12. Colocasia, ginger, turmeric and Zimikand are basically heavy feeders and therefore should be grown with supplemental does of nutrient. In addition to field crops, some short duration, less exhaustive and dwarf type inter-fillers like papaya, banana and guava can also be grown till these do not interfere with the main litchi crop.
13. Leguminous crops of like green gram, black gram, etc., cereals like wheat; oilseeds like mustard, sesame and groundnut and vegetables crops such as cabbage, cauliflower, tomato, potato, brinjal, cucumber, pumpkin, bitter gourd, tinda, lady's finger, etc. and spices like chillies can be successfully grown as intercrops in young litchi orchard.

Crop Combination. To intensify the orchard operation, a lot of work has been done on intercropping in orchard crops during initial years. Field crops like gram, mung, pea, potato, sweet potato, maize and vegetables like cauliflower, cabbage, spinach, okra, carrot, chillies, radish, onion, etc. can be grown successfully in the litchi orchard. Various other crops which can yield better under partial shade are elephant foot yam, sweet potato, ginger and turmeric.

Combinations of crops for inter spaces of litchi :

(*a*) Cabbage (Oct-Feb)—Cowpea (March-June)—Methi (July-Sept).

(*b*) Okra (Sept-Dec)—French bean (Jan-March)—Tomato (April-July).

(*c*) Potato (Nov-Feb)—Amaranth (March-May)—Cowpea (June-Aug) —Radish (Sept-Oct).

(*d*) Cauliflower (Aug-Nov)—Potato (Dec-March)—Dolicos (April-July).

(*e*) Cabbage (Sept-Dec)—Brinjal (Jan-May)—Cucumber (June-Aug.).

(*f*) Palak (Nov-Jan)—Cluster bean (Feb-May)—Brinjal (June-Oct.).

(*g*) Cow pea—Potato/Green gram—Niger/Black gram Mustard/Green gram/Black gram-gram/Cowpea Mustard/Cow pea—gram.

Fig. 5.6. Inter cropping model for a litchi orchard

VIII. Integrated Nutrient Management

Integrated crop management practices can be based on the principle that balanced fertilization in order to provide the appropriate allowance of nutrients to the crops, taking into account release from other sources such as organic manures, soil organic matter etc.

Application rates of either mineral or organic fertilisers shall be in accordance with international, national and local legislations.

- Nitrogen supply and timing shall be particularly well matched to meet crop demand as nitrogen leaching has significant environmental consequences.
- Optimum P and K balances shall be achieved as much as possible by rotational maintenance.
- Chopping and incorporation of crop residues as well as organic manure or compost shall be used to help improve soil fertility by increasing organic matter content, improving nutrient and water retention and reducing erosion.
- Manures and fertilisers shall be clearly marked and stored in a clean, dry location (preferably under cover), where there is no risk of contamination of watercourses,

- They shall not be applied to water logged, steep or frozen ground where there is a risk of run-off.
- Heavy metal build up shall be prevented by the use of good quality fertilizers. (Ref- Principles and Practices for Sustainable Fruit Production – Version 2009. SAI Platform Fruit Working Group. Pp. 1-14).

Fresh vs. composted manure. Fresh, non-composted manure will generally have a higher N content than composted manure. However, the use of composted manure will contribute more to the organic matter content of the soil. Fresh manure is high in soluble forms of N, which can lead to salt build-up and leaching losses if over applied. Fresh manure may contain high amounts of viable weed seeds, which can lead to weed problems. In addition, various pathogens such as *E. coli* may be present in fresh manure and can cause illness to individuals eating fresh produce unless proper precautions are taken.

Heat generated during the composting process will kill most weed seeds and pathogens. The microbially mediated composting process will lower the amount of soluble N forms by stabilizing the N in larger organic, humus-like compounds. A disadvantage of composting is that some of the ammonia-N will be lost as a gas. Compost alone also may not be able to supply adequate available nutrients, particularly N, during rapid growth phases of crops with high nutrient demands. The decision on what form of manure to use will ultimately depend on certification requirements, availability, and cost. Composted manure is usually more expensive than fresh or partially aged manure.

Heat-dried manure/compost. Drying manure or compost to low moisture content reduces their volume and weight, which lowers transportation costs, but it also requires energy inputs. Dried products can be easier to handle and apply uniformly to fields, especially those that have been processed into pellets. Heat drying also reduces pathogens if temperatures exceed 150 to 175°F for at least one hour and water content is reduced to 10 to 12 or less. The significant energy costs to heat-dry manure or compost at high temperatures are in contrast to the self-heating generated by microbial respiration during the composting process.

N availability from manure and compost. Bedding or litter will usually decrease the nutrient content of manure by dilution. If materials high in carbon (C) like straw or wood shavings are used as bedding, N availability may be reduced by the larger C/N ratio of the product. High C relative to N will lead to a tie-up of N, potentially causing N deficiency in the crop. A C/N ratio of 25/1 or greater will lead to N tie-up in the soil. A C/N ratio of less than 25/1 will release N to the crop. The C/N ratio is also an important consideration in the use of various composts, as well as a controlling factor in the composting process itself.

The N in manure or compost will not all be available to the crop the first year. In non-composted manure, some of the N will be lost to the atmosphere during application in the form of ammonia gas. Manure applied to land should be incorporated as soon as possible after application to avoid ammonia losses. Because the N is stabilized in organic compounds, atmospheric losses of ammonia are not as critical in compost applications, although incorporation is highly recommended. N in the organic form will need to be broken down by microorganisms before plants can use it. The release of N from manure or compost will depend on the type and age of the manure. Anywhere from 5 to 90 per cent of the organic N can be released to useable forms the first year. For dairy manure, 40 to 60 per cent of the N is generally considered to be available the first year, and 30 to 40 per cent of the N is available the second and third years. For poultry manure 50 to 75 per cent of the N is available the first year and 20 to 25 per cent is available the second and third years. For composted dairy manure, 5 to 20 per cent of the N is available the first year.

Use of soil and tissue tests, and monitoring of crop growth, will help in determining the amount of residual N, P, and other nutrients in the soil and the need for further applications.

Green manures/cover crops. Crops that are incorporated into the soil while still green are referred to as green manures. Cover crops are similar to green manures, but are usually grown to protect soil from erosion during the non-growing season. Because topsoil is higher in organic matter and nutrient content than subsoil, controlling erosion is an important method of conserving soil nutrients. Green manures and cover crops are both used to supply N and increase soil organic matter. Legumes such as clover and alfalfa can fix between 100 and 200 lbs of N per acre in one year. The use of grasses such as rye or oats without a legume will not increase the N content of the soil. These crops are used for increasing soil organic matter content. They can also scavenge residual N from the previous crop and keep it from being lost by leaching. Some green manure crops accumulate high levels of P and are thought to increase P availability to subsequent crops by returning it to the soil in organic form.

Other sources of N. Sources of N for organic production other than manures and composts include : blood meal, fish emulsion, fish protein, kelp/seaweed, animal tankage, and various vegetable meals. Municipal yard waste and vegetable/plant wastes from grocery stores, vegetable processing plants, and restaurants are other sources of organic material that can be recycled and used for crop production. In most situations these wastes should be composted before use to reduce odours, stabilize the nutrients, and facilitate ease of spreading. Sewage sludge is an organic fertilizer that can supply N and other nutrients, but sewage sludge is not acceptable for organic production.

Phosphorus and Potassium Management. The majority (70-90%) of P and K in manure will be available to the crop the first year of application. The P and K content of manure will vary with type of manure and storage. Poor storage of manure or compost will lead to losses of K through leaching. Have the product tested for P and K content before application. P sources other than manure for organic production include : bone meal, fish and poultry meal, and rock phosphate. K sources other than manure for organic production include: alfalfa meal, kelp meal, greensand, wood ash, langbeinite (sul-po-mag) and potassium sulphate.

Secondary and Micronutrients. Secondary (Ca, Mg, S) and micronutrient (Fe, Mn, copper, Zn, B, and Mo) needs will be satisfied in most situations through manure, compost, and liming amendments. A major advantage of organic nutrient sources is that they usually contain at least a small amount of all the essential plant nutrients. Deficiencies of some secondary and micronutrients may occur on susceptible crops under certain soil conditions conducive to deficiencies.

IX. Introduction of High Density Planting and Meadow orcharding

The minimum vertical distance between any two trees or plants is referred as the planting distance and this varies depending upon many factors. The principles in deciding the planting distances are the following.

1. Trees when fully grown, the fringes of trees should touch each other but the branches should not interlock.
2. Trees root will spread over a much larger area than top and there should be proper room for the roots to feed without competition.

Factors which decide the planting distance are the following :

1. Kind of fruit trees-mangoes are planted at a distance of 10 m × 10 m, guavas at a distance of 5 m × 5 m while papayas are planted at a distance of 2 m × 2 m.
2. Rainfall-wider spacing be given in low rainfall areas than the high rainfall areas.
3. Soil type and soil fertility-in heavy soils less spacing should be given because the top and root growth are limited.
4. Rootstocks-trees of the same variety grafted on different root stocks will grow to different sizes and as such require different planting distances. *e.g.*, Apple.
5. Pruning and training-trees trained on head system requires closer spacing than the other type of training system.
6. Irrigation system in general, if the spacing is too wide, the yield per unit area would be greatly reduced. If the trees are too close together, the trees grow tall rendering pruning, spraying and harvesting difficult. Close planting results in a greater yield per unit area in the early life of the tree but less in the more important later years. Close planting is therefore a false economy.

The total number of trees per hectare for various important horticultural crops under (*a*) square (*b*) hexagonal and (*c*) triangular system of planting are given below :

Crop	Planting distance (in m)	No. of trees per hectare		
		Square system	Hexagonal system	Triangular system
Mango	10 × 10	100	115	89
Sapota	8 × 8	156	118	139
Clove	6 × 6	277	320	248
Acid lime	5 × 5	400	461	357
Coconut	7.5 × 7.5	177	205	159

It may be seen that hexagonal accommodates 15 more number of plants while triangular system accommodates 11 lesser number of plants.

High density planting system. Planting of fruit trees rather at a closer spacing for obtaining maximum productivity per unit area without sacrificing quality is often referred as 'High density planting' or HDP. This technique was first established in apple in Europe during sixties. In this system, four planting densities are recognized for apples *viz.,* low HDP (< 250 trees/ha), moderate HDP (250-500 tree/ha), high HDP (500 to 1250 trees/ha) and ultra high HDP (>1250 trees/ha). Recently, super high density planting system has been also established in apple orchards with a plant population of 20,000 trees per ha. In some orchards, still closer, planting of apple trees is followed (say 70,000 trees/ha) which is often referred as 'meadow orchards'.

Advantages of HDP are :

(*i*) Early cropping and higher yields; the average yield in apple is about 5.0 t/ha under normal system of planting but about 140.0 t/ha under HDP

(*ii*) Reduced labour costs

(*iii*) Improved fruit quality

Characteristics of HDP are :

(*a*) The trees of HDP should have maximum number of fruiting branches and minimum number of structural branches.

(*b*) The trees are generally trained with a central leader surrounded by nearly horizontal fruiting branches.

(*c*) These branches should be so arranged and pruned in such a way that each branch casts a minimum amount of shade on other branches.

(*d*) The height should be one and half its diameter at the base. A key to successful HDP depends upon the control of tree size.

This is achieved by

(*a*) Use of size controlling root stocks. In apple, dwarfing root stocks and intermediate stocks like MM 106, MM 109, and MM 111 are used to control the size of the plant. In pears, Quince A, Adam and Quince-C are commonly used as dwarfing root stocks.

(*b*) Use of spur type scions-In temperate fruit crops like apple, the cultivars can be classified into a spur type or non-spur type. The spur types which have restricted annual growth are alone suitable for HDP.

(*c*) Training and pruning methods to induce dwarfness-under Indian conditions, apple trees trained under spindle bush, dwarf pyramid, cordon systems to contain the growth of the trees appreciably for HDP systems.

(*d*) Mechanical device and use of chemicals to control size-Growth regulators such as daminozide, ethephon, chlormaquat and paclobutrazal are extensively used to reduce shoot growth by 30 to 0. This results in increased flowering in the subsequent years and may be useful in encouraging earlier commercial fruit production in strongly vegetative fruitful young trees.

(*e*) Planting system for HDP : The success of HDP depends upon the right choice of planting system. Generally, rectangular planting with single, double and three row plantings are followed. In single row planting, the distance within the row is close, whereas the distance between the rows is wide (4 m × 2 m). In double row planting, a wider spacing is given after every two rows (4 + 2 × 2 m) whereas in three row planting, a wider spacing is given after every three rows (4 + 2 m × 2 m × 2 m). In meadow orchard system, a bed of 10 to 15 rows is closely planted *(nay* 30 × 45 cm) and separated by alleys of 2.5 m width between beds. This system is also called bed system (N. Kumar, 1997).

X. Plant protection measures by means of low inputs

Integrated pest and disease management (IPDM) is an approach to pest and disease management that utilises a range of techniques intended to reduce economic damage. There are several techniques you can use to manage pests and diseases.

Reduce the impact of pests by manipulating cultural practices. Some of the strategies employed include the following :

- Rotate crops when planting a new orchard-plant botanically unrelated crops. Avoid replanting old orchard sites with the same crop as there may be pest or disease carryover (for example, *Armillaria* in stone fruit and citrus orchards, or burrowing nematodes, *Radopholus similis*, in bananas).
- Cultivate the soil to disrupt the life-cycles of soil inhabiting pests, weeds and diseases. Minimise environmental damage by cultivating at the appropriate time and at the correct soil moisture. Keep the number of cultivations to a minimum and do not cultivate steep slopes.
- Vary the harvest time. Selective harvesting periods may avoid the worst effects of pests at the crop's most vulnerable period. You may need to select cultivars which mature at a different time to the pest's peak occurrence.
- Vary crop spacing. Spacing and siting of crops may affect the relative growth rate of a plant and the behaviour of pests in searching for food or egg laying sites. Closer plantings may favour the spread of beneficial species within the orchard, but may also reduce air movement, increasing levels of disease.
- Prune crops. Thinning, topping and pruning of dead or diseased material from fruit trees reduces disease incidence and maintains productivity and vigour.
- Keep orchards clean. Remove pest breeding and hibernating sites by removing, destroying or composting old or fallen fruit. Clean paddock borders to reduce pest migrations and weed infestations. Study the insects and wildlife inhabiting paddock borders first to avoid destroying habitats of beneficial species.

Bird netting or paper bags over fruit exclude flying foxes and birds (Robyn, 2008).

Natural Pesticides. Some plants contain components that are toxic to insects. When extracted from the plants and applied on infested crops, these components are called botanical pesticides or botanicals. The use of plant extracts to control pests is not new. Rotenone (*Derris* sp.), nicotine (tobacco), and pyrethrins (*Chrysanthemum* sp.) have been used widely both in small-scale subsistence farming as well as in commercial horticulture. Most botanical pesticides are contact, respiratory, or stomach poisons. Therefore, they are not very selective, but target a broad range of insects.

This means that even beneficial organisms can be affected. Yet the toxicity of botanical pesticides is usually not very high and their negative effects on beneficial organisms can be significantly reduced by selective application.

Preparation of Neem. The preparation and use of botanicals requires some know-how, but little material and infrastructures. It is a common practice under many traditional agricultural systems. Neem, for example, which is derived from the neem tree (*Azadiracta indica*) of arid tropical regions, contains several insecticidal compounds. The main active ingredient is *azadiractin*, which both deters and kills many species of caterpillars, thrips and whitefly. Both seeds and leaves can be used to prepare the neem solution. Neem seeds contain a higher amount of neem oil, but leaves are available all year. A neem solution loses its effectiveness within approximately 8 hours of preparation, and when exposed to direct sunlight. It is most effective to apply neem in the evening, directly after preparation, under humid conditions or when the plants and insects are damp. There are different recipes for the preparation of a neem solution. One is the Neem seed kernel extract: Pound 30 g neem kernels (that is the seed of which the seed coat has been removed) and mix it in 1 liter of water. Leave that overnight. The next morning, filter the solution through a fine cloth and use it immediately for spraying. It should not be further diluted.

Preparation of Pyrethrum. In the tropics, pyrethrum is grown in mountain areas because it needs cool temperatures to develop its flowers. Pyrethrins are insecticidal chemicals extracted from the dried pyrethrum flower. The flower heads are processed into a powder to make a dust. This dust can be used directly or infused into water to make a spray. Pyrethrins cause immediate paralysis to most insects. Human allergic reactions are common. Pyrethrins break down very quickly in sunlight so they should be stored in darkness. Liquid formulations are stable in storage but powders may lose up to 20 per cent of their effectiveness in one year. Attention : Pyrethroids are synthetic insecticides based on pyrethrins, but more toxic and longer lasting. They are not allowed in organic farming.

Apart from plant extraction, there are some other natural pesticides that are allowed in organic farming. Although some of these products have limited selectivity and are not fully biodegradable, there are situations when their use is justified. However, in most cases, the desired effect is best reached in combination with preventive crop protection methods :

Sulphur : against fungal disease

Copper : against fungal disease (accumulates in the soil and harms soil organisms)

Sulphuric acidic argillaceous earth : against fungal disease

Ashes : against soil-borne disease

Slaked lime : against soil-borne diseases

Clay : against fungal diseases

Baking soda : against fungal diseases

Soft soap solutions : against aphids and other sucking insects

Light mineral oil : against various insect pests (harms natural enemies)

Sulphur : against spider mites (harms natural enemies)

Plant ashes : against ants, leaf miners, stem borers etc.

A major distinction between pest management in perennial fruit crops and in annual crops is that crop rotation is not an option (strawberries and to a lesser extent, brambles, are possible exceptions). The long term nature of fruit growing allows for the possible build up of a pest population over time. Conversely, it is also possible for such stable agricultural environments to sustain populations of beneficial organisms.

Though it is sometimes overstated, maintaining the plant in general health and good vigour is important in pest management. For fruit plants, this adage is more applicable to indirect pests (those pests that feed on foliage, stems etc.) than to direct pests (pests that feed on the fruit). For instance, an apparently healthy plum tree may set a good crop of fruit, yet lose it all to the plum. On the other hand, the same tree might suffer significant defoliation by caterpillars early in the season; yet, if it is in good vigour, it can compensate and bounce back quickly still producing a marketable crop that year. There are some cases where general plant health and freedom from stress does impart a form of "resistance" not technically genetic resistance to certain pests. Two examples are apple trees in good vigour actually smothering with sap or casting out invading flathead apple tree borers and plants not suffering drought stress being much less attractive to grasshoppers.

XI. Weed management

A well managed organic orchard should not develop a significant weed problem. A major objective in organic farming is to change the composition of the weed community so that the orchard gains maximum benefit. Sometimes, however, one species may dominate or a noxious weed (one that, by law, must be controlled) may be present, so the weed should be managed. Weed management around young trees and vines, is essential.

Mulching. Mulching helps to retain moisture, minimises weeds and improves biological activity in the soil. Inorganic and organic materials may be used for mulching. The mulch must be applied thickly enough to suppress the growth of weeds and germination of their seeds. A layer 2-3 cm deep should be adequate.

Grazing. Animals in the orchard can offer good weed control and provide additional benefits of extra income, added manure, insect control and the recycling of crop wastes. Chinese or 'weeder' geese and even carefully managed sheep are some examples. However, care should be taken, particularly during droughts, when livestock may feed on trees or vines.

It is a good idea to introduce stock into the orchard when they are young to develop their 'taste' for weeds and a grazing routine. Various systems of electric fencing are available to facilitate grazing (Robyn, 2008).

Crop Rotation. Rotation of crops is the most efficient measure to regulate seed and root weeds in annual crops. Changing the conditions of the crop interrupts the living conditions of the weeds thus inhibiting their growth and spread.

Sowing Time and Density. Weed pressure during the critical period (youth stage of the crop) can be reduced by choosing an optimal sowing time and by increasing sowing density.

Balanced Fertilization. It can support an ideal growth of the crop, which promotes the growth of the crop over the weeds.

Soil Cultivation. Weed cures before sowing, apply superficial stubble treatment works against persisting weeds under dry weather conditions. Minimum-tillage systems can increase the weed pressure.

A good organic weed management plan should present a minimum erosion risk, provide a "platform" for the movement of farm equipment, not impact adversely on pest management or soil fertility, while minimizing weed competition for water and nutrients. This philosophy has already been demonstrated in discussions regarding three effective weed control tools : cover crops, mulches and soil solarisation.

XII. Proper intercultural operation for higher yield of fruits

The guava plant grows symmetrically dome-shaped with broad, spreading, low-branching canopy and a shallow rooted small tree of 3 to 10 m in height, branching close to the ground and often heavily suckering from the base of the trunk. A single trunk tree is developed by proper pruning and training. The young guava tree is usually trained 3 to 4 months after field planting. Pruning begins at an early stage of plant growth to develop single trunk trees with well-spaced branches. Bearing guava need constant training and pruning to provide the desired tree shape for ease of management and for the health of the tree and to maximize fruit production. Larger fruit are primarily produced on vigorous shoots of two to three years of age. Most guava trees, whether propagated from seed or grafts, produce an abundance of suckers which should be removed from the trunk up to 50 cm above ground. A framework of four branches representing four quarters of the tree should be established. The crotch angles between the branches and the main stem should be wide enough to facilitate adequate light penetration and provide physical strength to support fruit load at maturity. A flat tree shape reduces labour for orchard management such as fruit thinning, fruit bagging and harvesting.

XIII. Proper labour management

Labour management is one of the major key among low inputs or to reduce the cost of cultivation for higher economic returns in fruit crop cultivation. The following criteria should be followed :

- Discrimination on the basis of ethnic groups, national origin, religion, disability, gender.
- Same rights and obligations shall be conceded to women and men.
- Decent working conditions and dignity shall be provided to all workers regardless of their employment status.

XIV. Improvement of farmer's technical know how

Support the training of fruit growers on all aspects of sustainable cultivation practices. They shall be provided with basic training and be competent on the following items: environmental protection; human health and hygiene; correct handling and use of substances or materials that are hazardous or harmful for human health; etc.

- In addition, the orientation of seasonal workers shall be promoted on issues concerning health, hygiene and protection of the environment.
- Literacy of all individuals at the farm who are not able to read and write shall be promoted.
- Producers shall strive to receive regular external advice, training and capacity building on the various aspects of farming in order to be able to increase the success of his activity over time.
- Employees and workers shall receive basic training in their area of work, especially if this involves handling of chemicals that can be hazardous to health, or if they work in the processing plant (Principles and Practices for Sustainable Fruit Production—Version 2009).

REFERENCES

Abay, F., Bjornstad, A. (2009). Specific adaptation of barley varieties in different locations in Ethiopia. *Euphytica,* **167** : 181-195.

Ceccarelli, S. (1994). Specific adaptation and breeding for marginal conditions. *Euphytic,* **77** : 205-219.

Dalgaard. (2001). "A model for fossil energy use in Danish agriculture used to compare organic and conventional farming". *Agri, Ecosystems and Envir.* **87** : 51-65.

David Norman. (1997). Defining and Implementing Sustainable Agriculture (Kansas Sustainable Agriculture Series, Paper #1; Manhattan KS : Kansas Agricultural Experiment Station, 1997). Available at K-State Research and Extension Web Site for Sustainable Agriculture: http://www.kansassustainableag.org/lIBRARY/ksas1.htm(8/23/07)

Dubrovsky, N.M., Burow, K.R., Clark, G.M., Gronber, J.M. and Hamilton, P.A. (2010). The quality of our nation's waters : Nutrients in the nation's streams and groundwater, 1992–2004. Circular 1350. Reston, VA: U.S. Geological Survey. http://pubs.usgs.gov/circ/1350/.

Eyhorn, F., Mader, P. and Ramakrishnan, M. (2005). The Impact of Organic Cotton Farming on the Livelihoods of Smallholders. *FIBL Research Report, October 2005.*

Foley, J.A., Ramankutty, N., Brauman, K.A., Cassidy, E.S. and Gerber, J.S. (2011). Solutions for a cultivated planet. *Nature* **478** : 337-342. doi: 10.1038/nature10452.

Haas, G., Geier, U., Frieben, B. and Kopke, U. (2005). Estimation of environmental impact of conversion to organic agriculture in Hamburg using the Life-Cycle-Assessment method, *Organic Agro-expertise Consultancy (www.agroexpertise.de), Germany.*

Jennifer, R., Brent, B., Diane, A. and Corey, R. (2009). Developing Organic and Reduced-Input Tree-Fruit Production Options for the Wasatch Front. Poster Presentations on Monday (2009-04-13). Utah State University. Coupling Humans and Complex Ecological Landscapes. 2009 US-IALE Symposium Snowbird, Utah | April 12-16, 2009.

Parr, J.F. (1990). "Sustainable Agriculture in the United States," In Sustainable Agricultural Systems, ed. by Clive A. Edwards, *et al.* (Ankeny IA: Soil and Water Conservation Society, 1990), p. 52. NAL Call # S494.5 S86S86.

Kumar, N. (1997). Introduction to Horticulture. Rajalakshmi Publications, 28/5-693, Vepamoodu Junction, Nagercoil. pp : 15.47-15.50.

Lal, R. (1989). Conservation tillage for sustainable agriculture: tropics v temperate environments. *Advances in Agro.* **42** : 186-197.

MacDonald, D.M. (2004). Agri Impact Assessment Study for Organic Cotton Farmers of Kutch and Surendranagar. *Agrocel Industries Ltd. September 2004.*

Murphy, K., Lammer, D., Lyon, S., Carter, B. and Jones, S.S. (2005). Breeding for organic and low-input farming systems : An evolutionary-participatory breeding method for inbred cereal grains. *Renew. Agric. Food Syst.* **20** : 48-55.

Principles and Practices for Sustainable Fruit Production-Version (2009). SAI Platform Fruit Working Group. pp- 1-14

Reganold, J.P., Elliot, L.F. and Unger, Y.L. (1987). Long-term effects of organic and conventional farming on soil erosion. *Nature.* **330** : 370-372.

Reganold, Glover, J.D., Andrews, P.K. and Hinman, H.R. (2001). "Sustainabilityof three apple production systems". *Nature* 410 (6831) : 926930. Doi:10.1038/35073574.PMID 11309616.

Robyn Neeson. 2008. Organic fruit production. Farming Systems Research, Yanco. *Primefact* 805. Pp 1-5.

Rosen Carl, J. and Bierman Peter, M. (2005). Maintaining Soil Fertility in an Organic Fruit and Vegetable Crops System. Department of Soil, Water, and Climate University of Minnesota University of Minnesota. M1191

Sanchez-Giron, V., Serrano, A., Hernanz, J.L. and Navarrete, L. (2004). Economic assessment of three long-term tillage systems for rainfed cereal and legume production in semi-arid central Spain. *Soil & Tillage Research*, **78**(1) : 35-44.

Shiva, V. (2004). Principles of Organic farming: Renewing the Earth's Harvest, *Navadanya Publications, New Delhi.*

Sthapit, B., Rana, R., Eyzaguirre, P. and Jarvis, D. (2008). The value of plant genetic diversity to resource-poor farmers in Nepal and Vietnam. *Int. J. Agric. Sustain.* **6** : 148-166.

United Nations Conference on Trade and Development (2003). Organic Fruit and Vegetables from the Tropics. Market, Certification and Production Information for Producers and International Trading Companies. UNCTAD/DITC/COM/2003/2. Pp-1-23.

USDA National Organic Program, see pages 45-46 and section 205.203.

Weibel, F. Suter, J.L. and Monney, P. (2006). Improved organic and low-input apple production using rootstocks tolerant to weed competition. ISHS Acta Horticulturae 772 : XXVII International Horticultural Congress-IHC 2006 : International Symposium on Enhancing Economic and Environmental Sustainability of Fruit Production in a Global Economy.

Yadav, R.K., Yadav, D.S., Rai, N. and Sanwal, S.K. (1999). Soil and water conservation through horticultural intervention in hilly areas. *ENVIS Bulletin,* **14**(1) : Himalayan Ecology.

●●●

6

Eco-Friendly Soil Management Approaches to Optimize Horticultural Production

SANJAY ARORA AND ANIL SHARMA

India is endowed with varying ecological situations and it is possible to grow almost all groups of Horticultural Crops. India produces 45.5 million tones of fruits annually from an area of 3.8 million ha, resulting in 88 g per capita availability daily as against a recommended consumption of 120 g per capita/day. Besides, it's productivity of 12.0 t/ha is also very low when compared to many countries. There are several constraints in increasing the area under horticultural crops and their production. Yet to sustain the production and increase the productivity of these crops, judicious nutrition management is one of the options. This can be possible not only through use of chemical fertilizers but also organic sources as well as use of certain plant growth promoting microbes. Horticulture is fast emerging as a major commercial venture, because of higher remuneration per unit area and the realization that consumption of fruits and vegetables is essential for health and nutrition. In the last one decade, export potential of horticultural crops has significantly increased attracting even multinationals into floriculture, processing and value added products. Since the horticultural produce especially fruits and vegetables are consumed afresh, consumers expect residue-free produce. In modern society where consumers are becoming increasingly health conscious and environmentally aware, a major market for organic foods has developed. The organic sector, in particular, has sprung back into life to become one of the most dynamic sectors in the international food market.

SOIL FERTILITY SCENARIO IN INDIA

Indian agriculture is operating under the pressure of multi nutrient deficiencies. Nitrogen deficiency in soils is almost universal. Phosphorus deficiencies are not far behind those of N as in 95 per cent districts, the P fertility is either low or medium. It is known that when exchangeable K is not rapidly replenished, crops start drawing on the non-exchangeable K, resulting in soil mining and depletion of soil K reserves (Tiwari and Rao, 2005). The K deficient areas have increased and crops in many areas are responding to K where they were not responding some years ago. Crops in general remove as much or more K than they remove N but average consumption of K_2O ha^{-1} is still 8.7 kg while that of N is 58.7 kg ha^{-1} at a highly unbalanced N : K_2O ratio of 7 : 1.

With intensive cultivation of high yielding varieties, deficiencies of secondary and macronutrients caused declining productivity of crops in many soils. Among secondary nutrients, sulphur deficiency is an important problem in many states and soils in 130 districts are considered to be suffering from S deficiency to varying extent.

Though it is believed that magnesium deficiency occurs in acid soils or where exchangeable Mg is below 1 meq per 100 g or less than 10 per cent of soil CEC is occupied by Mg^{++} but the problem seems to be of greater magnitude than expected. Among micronutrients zinc, boron and iron deficiencies were found most common problem. Soil application of zinc, boron and sulphur was found more effective than their foliar application and *vice-versa* for iron and manganese.

ORGANIC INPUTS IN SOIL FERTILITY MANAGEMENT

The organic inputs to soil (animal manure and composts, crop residues, green manures, urban wastes etc.) are known to have favorable effects on soil physical, chemical and biological processes and its overall health. Organic materials generally improve the soil organic matter (SOM), a basic indicator of soil health and resilience, in addition to supply of substantial quantities of plant nutrients to enrich soil fertility.

Though abundant quantities of organics are generated annually, the entire quantity is not available for returning to soils because of several competing alternate uses in rural India. Tandon (1997) estimated organic resources availability to be 307 Mt in 2025 in India (Table 6.1). If used properly, these quantities of organic resources can supply plant nutrients ($N + P_2O_5 + K_2O$) to the tune of 7.75 million tones in 2025, and certainly help minimize the negative nutrient balances.

Addition of organics enhances the soil microbial population. Chowksey (1991) reported a decline in the count of bacteria and actinomycetes with inorganic fertilizer application. The counts of all the microbes were substantially increased when inorganics was combined with organics (Table 6.2).

TABLE 6.1

Some projections on the tappable* organic resources and their nutrient supply ($N + P_2O_5 + K_2O$) for agriculture in India during 2025

Resource	Tappable quantity (mt)	Nutrient supply (mt)
Human excreta	17	2.10
Livestock dung	128	2.26
Crop residues	160	3.39
Total	307	7.75

* Tappable = 80% of excreta, 30% of dung and 33% of crop residues *(Source : Tandon, 1997)*

TABLE 6.2

Effect of INM on viable microbial population in soil Microbial population (number/g of soil)

Treatments	Bacteria ($\times 10^9$)	Actinomycetes ($\times 10^6$)	Fungi ($\times 10^6$)	Rhizobia ($\times 10^6$)
No fertilizer application	32.08	2.90	0.17	16
NPK 20 : 80 : 20 kg/ha	21.60	1.40	0.53	104
NPK 20 : 80 : 20 FYM 15 t/ ha	95.25	6.20	0.83	150

INTEGRATED NUTRIENT MANAGEMENT

Integrated Plant Nutrient Supply (IPNS) system refers to the maintenance or adjustment of soil fertility and plant nutrient supply at an optimum level for sustaining the desired productivity through optimization of the benefits from all possible sources of plant nutrients in an integrated manner (Roy, 1995). Conceptually, the IPNS strives to achieve : (*a*) regulated nutrients supply for optimum crop growth and productivity, (*b*) maintenance or some times an improvement in soil fertility, and (*c*) minimum adverse impact on agro-ecosystem quality by means of striking a balance among various nutrient sources *viz.*, soil fertilizers, organic manures and bio-inoculants (Fig. 6.1).

It is well recognized that among all the cultivated crops species, banana removes extremely high amounts of potassium from the soil. Very high amounts of potassium are removed by such fruit crops as citrus and pineapple. Tuber crops like potato, cassava and sweet potato have also high K requirements as do a number of vegetable crops such as cabbage and cauliflower. The root and tuber crops are notable in that they need to take up large amounts of potassium most of which is removed from soils in the roots and tubers.

On the whole, the above indicated horticultural crops uptake more potassium than nitrogen or phosphorus. Thus, an adequate fertilization with potassium is a major precondition for obtaining high yields of these crops.

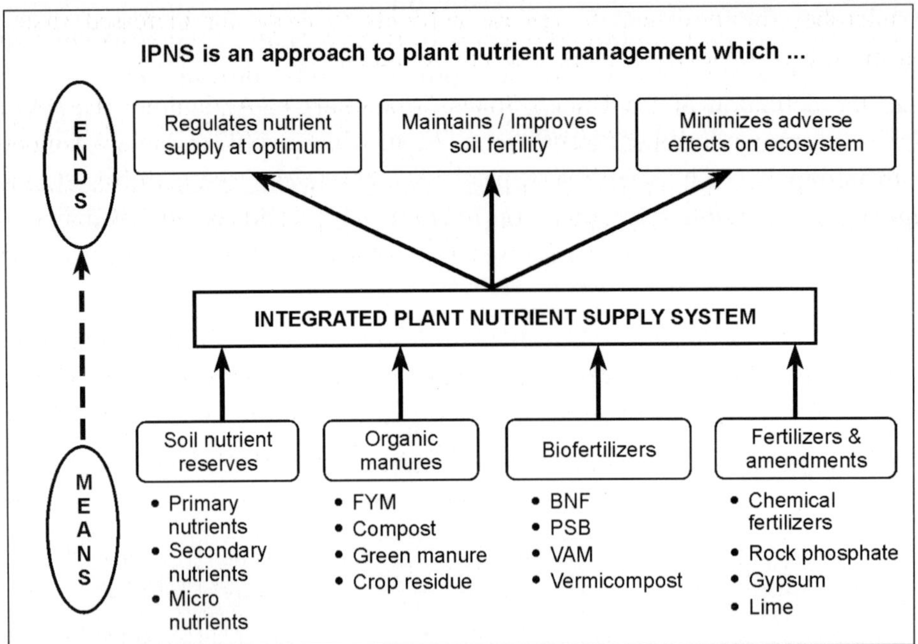

Fig. 6.1. The concept of Integrated Plant Nutrient Supply

(***Source*** : Subba Rao and Damodar Reddy, 2005)

Among the major nutrients, potassium not only improves yields but also benefits various aspects of crop quality. Below are summarized the most important benefits from potash fertilizer application, originated from the key role potassium plays in the growth and development of plants :

- increase in crop yield;
- improvement in both the appearance and marketability of crop production (*i.e.* size, shape, colour, hygienic properties);
- enhancement of nutritional value of crop production (*i.e.* the content of vitamin C and antioxidants, starch content in potato);
- decrease in both the incidence and severity of pest and diseases attacks;
- improvement of plant resistance to drought conditions;
- decrease of storage losses, enhancement of shipping quality and extending of shelf life of crop production;
- improvement in the processing quality of crop production.

WHAT IS ORGANIC FARMING ?

Organic farming system in India is not new and is being followed from ancient time. It is a method of farming system which primarily aimed at cultivating the land and raising crops in such a way, as to keep the soil alive and in good health by use of organic wastes (crop, animal and farm wastes, aquatic wastes) and other biological materials along with beneficial microbes (biofertilizers) to release nutrients to crops for increased sustainable production in an eco friendly pollution free environment.

As per the definition of the United States Department of Agriculture (USDA) study team on organic farming "organic farming is a system which avoids or largely excludes the use of synthetic inputs (such as fertilizers, pesticides, hormones, feed additives etc.) and to the maximum extent feasible rely upon crop rotations, crop residues, animal manures, off-farm organic waste, mineral grade rock additives and biological system of nutrient mobilization and plant protection".

FAO suggested that "Organic agriculture is a unique production management system which promotes and enhances agro-ecosystem health, including biodiversity, biological cycles and soil biological activity, and this is accomplished by using on-farm agronomic, biological and mechanical methods in exclusion of all synthetic off-farm inputs".

Fig. 6.2

Need of organic farming

With the increase in population our compulsion would be not only to stabilize agricultural production but to increase it further in sustainable manner. The scientists have realized that the 'Green Revolution' with high input use has reached a plateau and is now sustained with diminishing return of falling dividends. Thus, a natural balance needs to be maintained at all cost for existence of life and property. The obvious choice for that would be more relevant in the present era, when these agrochemicals which are produced from fossil fuel and are not renewable and are diminishing in availability. It may also cost heavily on our foreign exchange in future.

The key characteristics of organic farming include

1. Protecting the long term fertility of soils by maintaining organic matter levels, encouraging soil biological activity, and careful mechanical intervention.
2. Providing crop nutrients indirectly using relatively insoluble nutrient sources which are made available to the plant by the action of soil micro-organisms.
3. Nitrogen self-sufficiency through the use of legumes and biological nitrogen fixation, as well as effective recycling of organic materials including crop residues and livestock manures.
4. Weed, disease and pest control relying primarily on crop rotations, natural predators, diversity, organic manuring, resistant varieties and limited (preferably minimal) thermal, biological and chemical intervention.
5. The extensive management of livestock, paying full regard to their evolutionary adaptations, behavioural needs and animal welfare issues with respect to nutrition, housing, health, breeding and rearing.
6. Careful attention to the impact of the farming system on the wider environment and the conservation of wildlife and natural habitats.

Organic Cultivation of Fruits

Organic fruit production essentially excludes the use of many inputs associated with modern farming, most notably synthetic pesticides and fertilizers. Chemical fertilizers are commonly used be-cause they can ensure quick avail-ability of the nutrients to the plants. However, continuous application of inorganic fertilizers deteriorates the soil's health (Warade *et al.,* 1995). Therefore, chemical fertilizers must be integrated with sources of organic manure which are eco-friendly for sustaining productivity without deteriorating effects on soil health and the environment. The yield can be increased or sustained, but the quality is improved through the balanced application of organic and inorganic fertilizers. To the maximum extent possible, organic farming systems rely upon crop rotations, crop residues, animal manures, legumes, green manures, off-farm organic wastes, mechanical cultivation, mineral-bearing rock powders and biological pest control. These components maintain soil productivity and tilth, supply plant nutrients and help to control insects, weeds and other pests.

CULTURAL PRACTICES

Site selection

All factors regarding site suitability for conventional fruit plantings (air and water drainage, etc.) apply even more and so to organic operations. While conventional farmers may fall back on chemical fertilizers and pesticides to compensate for poor site decisions, organic farmers largely give up such luxuries. The presence of certain weeds and forage species are also of particular concern to the organic farmer. Bermuda grass, Johnson grass and several other species can be quite problematic to farmers and are difficult to control through nonchemical means.

Growing fruit crops offers an advantage to farmers interested in sustainable agriculture. Because fruit plantings are perennial, the soil may not require additional tillage or cultivation beyond that needed at establishment, thereby minimizing soil erosion. Because the potential for erosion is low, hillsides and other sites unsuitable for tillage agriculture can safely and successfully produce fruit crops.

Crop selection

Environmental constraints (climate, presence of certain pests and diseases, suitable soils, etc.) can greatly impact the suitability of a given site or even a bioregion for organic production of a given fruit crop. Generally speaking, some perennial fruits are easier to grow organically than others. The small fruits (berries) for example, seem easier to grow organically than the tree fruits in almost all locations. Lastly, successful organic fruit growing may depend largely on whether the venture is for home production or for commercial sales.

Site preparation

Attention to the details of site preparation can go a long way towards reducing weed and disease problems and assuring a vital planting through soil improvement.

In general, fruit crops do not require highly fertile soils for good production. In fact, highly fertile soils, rich in nitrogen, can promote too much vegetative growth at the expense of fruiting. A nutritionally balanced soil, proper oil pH and plentiful amounts of organic matter are the fundamentals of an organic fertility management plan for fruits. Pre plant soil improvement for organic fruit plantings is typically accomplished through some combination of cover cropping/green manuring and the use of imported materials, which may include manures, compost, rock powders and organic wastes.

Of particular importance at this stage is the application of required amounts of lime or sulphur. The need for lime or sulphur is dependent on the crop to be grown and the results of soil testing. Adjusting the oil pH with lime (to raise the pH) or sulphur (to lower pH) is much more easily done before planting. Most fruit plants perform best around pH 6.5, although they tolerate a pH range between 5.5 and 7.2. Blueberries are an exception. They require an acid oil-ideally pH 4.8 to 5.2 soil testing should also be used to guide applications of manures and other rock powders to avoid nutrient imbalances.

Selection of specific cover crops and their management varies with location, depending on such factors as seasonal rainfall, soil type, soil erosion potential, available equipment and seed cost. For example, *Crotalaria* probably performed less effectively than *Sesbania* at different centre because it is poorly adapted to low, wet soils. Some other warm season cover crops that might be considered include aggressive maize varieties and forage soybean varieties.

Crops such as berries (strawberries) are often grown on raised beds to encourage drainage and reduce root rot disease problems.

Soil solarization

Another technique for site preparation is soil solarization. The process involves placing transparent polyethylene plastic on moist soil during the hot summer months to increase soil temperatures to levels lethal to many pests. Solarization suppresses weeds and eliminates many potential soil disease and nematode problems. Soil solarization is somewhat expensive and is not frequently used to prepare sites for perennial fruits, with the possible exception of strawberries.

Orchard Floor Management and Mulching

The floor or inter row areas of perennial fruit plantings may be managed in a variety of ways, ranging from clean cultivation to various combinations of cover crops and organic mulches. In terms of controlling erosion at least, a system that results in full or close-to-full ground coverage is best. If the decision is made to plant a between row ground cover, it should consist of a species adapted to the region and to the fruit farmer's management plan. Where adapted, orchard grass and other cool season grasses are often recommended because they go dormant during the heat of the summer, thereby minimizing competition with the fruit crop for water. With proper fertility management these grasses may also provide plentiful material for mulch. Many warm season legumes are deep rooted and compete with the trees for water, they should not be allowed to grow under the tree canopy. However, legumes and the mulch made from leguminous ground covers can provide significant nitrogen to fruit trees or vines. Furthermore, even though they complete with the fruit plants for water, legumes (like the grasses) increase the water permeability of the soil. Increasing the organic matter content of the soil also increases the soil's moisture retentiveness.

Many researchers have also investigated the relationship between orchard floor management and pest control. In some place apple orchards peach orchards, a diverse mix of cover crop spices provided habitat and food for an array of beneficial organisms, resulting in a decrease of orchard pests.

After a planting is established, mulching with organic materials such as straw, leaves or sawdust can provide significant weed suppression. If applied thickly enough or supplemented with sheets of paper or cardboard as the bottom layer, complete suppression might be achieved. Application of mulch varies somewhat by crop. In strawberries for instance, mulch might cover entire aisles between rows or only be placed next to the bed to inhibit encroachment by creeping weeds. In vineyards and orchards, mulch might be placed only under individual trees or vines or along the entire row, ideally extending to the drip line.

Generally, mulch should be kept well away from the trunk to reduce damage from voles. This is especially important in winter. Keeping the mulch 8-12 inches away from the trunk also reduces the likelihood of crown rot and other diseases in susceptible species most notably applies on certain rootstocks. In mulberries, sawdust mulch is commonly spread along the entire row with extra sawdust mounded around the canes, often to a depth of 8 or more inches. In addition to controlling weeds, mulching with organic materials improves the soil by enhancing soil aggregation and water availability.

Varietal selection

Because the plants are perennial and represent a considerable investment in both time and money, it is important to start the fruit planting with the optimum varieties for location and markets.

Good information on varietal selection is available from Cooperative Extension, nurseries and local commercial farmers. It is also important to obtain clean planting stock. Buying from reputable nurseries that provide stock certified by certifying inspectors to be free of diseases and insect pests is recommended.

Genetic resistance refers to inheritable traits in the plants that inhibit disease and pest damage. Choosing genetically resistant cultivars is a very important control measure for organic farmers especially with regard to disease management. In some cases, such as bacterial spot in peaches, cultivar resistance is the best or only control.

Aphid resistant berries, *Phylloxera* resistant grape rootstocks, wooly aphid resistant apple rootstocks, mite resistant strawberry cultivars, and nematode resistant peach rootstocks are available. As important as this resistance is there is no cultivar of any fruit species with multiple insect pest resistance; therefore, means other than resistance will most likely have to be employed to protect fruit plant from a complex of several pest species.

ORGANIC FERTILIZATION PRACTICES

Harvested fruit, being largely water, removes relatively few nutrients from the soil, compared to other crops. Therefore, a significant amount of the fertility needs of fruit crops can be met through cover crop management and organic mulches in systems which use them and by the application of lime and other slow release rock powders at the preplant stage. Still, supplementary fertilization is often required for optimum growth and production. Some information useful to planning a supplementary fertilization program for perennial fruits include :

- Organic farmers generally employ relatively non-soluble fertilizer materials, such as compost, manures, plant derived byproducts like cottonseed meal or animal byproducts like feather or blood meals. To insure adequate decomposition and timely nutrient release of these slowly available materials, early spring application is encouraged.

Early application also reduces the tendency towards late season growth, which may result in winter damage.

- Surface application of organic fertilizer materials without incorporation is sometimes wasteful of the nitrogen contained in those materials. This is especially true of manures. However, incorporation by tillage could damage the roots of the fruit plants and increase erosion.
- To bypass some of the "problems" associated with slowly available organic materials, some organic fruit farmers choose soluble organic fertilizers such as fish emulsion, soluble fish powder or water soluble blood meal in some cases applying these as a foliar spray. Since these are relatively expensive, the prudent farmer seldom relies exclusively on such materials.
- Most organic fertilization programs focus on supplementing nitrogen as the key element since it is needed in the greatest amount for the crop. One way to determine a proper application rate for an organic fertilizer is to obtain a conventional chemical recommendation for the planting and calculate the approximate amount of an organic fertilizer required to meet it. For instance, if the recommendation for a berry crop is 160 kg of actual nitrogen per acre, a farmer choosing to use cottonseed meal (approximately 7% N) would need to apply about 2000 kg of that material to each acre. Such calculations are simplistic however and may lead to spending more on fertilizer inputs than in necessary. One reason is because biologically healthy soils fix and release greater amounts of nitrogen naturally than those which are not.
- Further more, when making fertilizer calculations based on nitrogen, the farmer needs to credit the estimated contributions made by legume cover crops and or leguminous mulches, where these are used. A cover crop of subterranean clover, properly fertilized and inoculated, can fix from 200 to 1000 kg of nitrogen per acre annually in a living mulch system.
- Basing application rates solely on nitrogen content can cause problems when the fertilizers themselves are imbalanced. Repeated use of poultry manure for example which is very high in phosphate, can lead to pollution problems and a zinc deficiency in the crop. These problems can be averted by regular soil testing and adjusting fertilizer selection and rates accordingly.
- The most reliable means for determining whether fertilization is adequate is to combine field observations with soil and tissue testing. Poor yields, unusual colouration of leaves, and poor plant growth are all clues to a possible nutritional imbalance or deficiency. For example, less than 10 inches annual elongation of the branches of most fruit trees probably indicates a nitrogen deficiency. A corky bark on certain apple varieties may indicate an over availability of manganese in the soil.

A foliar analysis measures the nutrient content of the leaves and can identify a deficiency or excess well in advance of visible symptoms. It is more helpful than a soil test because the foliar analysis is a measure of what the plant is actually taking up, while soil analysis only measures what is in the soil, which may or may not be available to the plant. Annual foliar analyses generally provide the best guide for adjusting supplementary nitrogen fertilization.

BIOFERTILIZERS FOR FRUIT CROPS

Biofertilizers are defined as preparations containing living cells or latent cells of efficient strains of microorganisms that help crop plants' uptake of nutrients by their interactions in the rhizosphere when applied through seed or soil. They accelerate certain microbial processes in the soil which augment the extent of availability of nutrients in a form easily assimilated by plants.

Biofertilizer is a natural product carrying living microorganisms derived from the root or cultivated soil. So they don't have any ill effect on soil health and environment. Besides their role in atmospheric nitrogen fixation and phosphorous solubilisation, these also help in stimulating the plant growth hormones providing better nutrient uptake and increased tolerance towards drought and moisture stress. A small dose of biofertilizer is sufficient to produce desirable results because each gram of carrier of biofertilizers contains at least 10 million viable cells of a specific strain (Anandaraj and Delapierre, 2010).

Azotobacter and *Azospirillum* are the two most important non-symbiotic N-fixing bacteria in non-leguminous crops. Under appropriate conditions, *Azotobacter* and *Azospirillum* can enhance plant development and promote the yield of several agricultural important crops in different soils and climatic regions (Okon and Labendera-Gonzalez, 1994). These beneficial effects of *Azotobacter* and *Azospirillum* on plants are attributed mainly to an improvement in root development, an increase in the rate of water and mineral uptake by roots, displacement of fungi and plant pathogenic bacteria and, to a lesser extent, biological nitrogen fixation (Okon and Itzigshohn, 1995). Besides N_2 fixation, *Azotobacter* synthesizes and secretes considerable amounts of biologically active substances like B vitamins, nicotinic acid, pantothenic acid, biotin, heteroxins, gibberelins etc. which enhance root growth of plants (Rao, 1986). Another important characteristic of *Azotobacter* association with crop improvement is secretion of ammonia in the rhizosphere in the presence of root exudates, which helps in modification of nutrient uptake by the plants (Narula and Gupta, 1986). The ability of *Azospirillum* to produce plant growth regulatory substances along with N_2 fixation stimulate plant growth and thereby productivity. The changes that occur in the plant roots help in transport of minerals and water (Sarig *et al.*, 1988).

Very often microorganisms are not as efficient in natural surroundings as one would expect them to be and therefore artificially multiplied cultures of efficient selected microorganisms play a vital role in accelerating the microbial processes in soil.

Use of biofertilizers is one of the important components of integrated nutrient management, as they are cost effective and renewable source of plant nutrients to supplement the chemical fertilizers for sustainable agriculture. Several microorganisms and their association with crop plants are being exploited in the production of biofertilizers. They can be grouped in different ways based on their nature and function.

S.No.	Groups	Examples
Nitrogen fixing Biofertilizers		
1.	Free-living	*Azotobacter, Beijerinkia, Clostridium, Klebsiella, Anabaena, Nostoc,*
2.	Symbiotic	*Rhizobium, Frankia, Anabaena azollae*
3.	Associative Symbiotic	*Azospirillum*
Phosphate Solubilizing Biofertilizers		
1.	Bacteria	*Bacillus megaterium* var. *phosphaticum, Bacillus subtilis Bacillus circulans, Pseudomonas striata*
2.	Fungi	*Penicillium sp. Aspergillus awamori*
Phosphate Mobilizing Biofertilizers		
1.	Arbuscular mycorrhiza or AM	*Glomus* sp., *Gigaspora sp., Acaulospora* sp., *Scutellospora* sp. and *Sclerocystis* sp.
2.	Ectomycorrhiza	*Laccaria* sp., *Pisolithus* sp., *Boletus* sp., *Amanita* sp.
3.	Ericoid mycorrhizae	*Pezizella ericae*
4.	Orchid mycorrhiza	*Rhizoctonia solani*
Biofertilizers for Micro nutrients		
1.	Silicate and Zinc solubilizers	*Bacillus* sp.
Plant Growth Promoting Rhizobacteria		
1.	Pseudomonas	*Pseudomonas fluorescens*

Nitrogen fixing bacteria and phosphate solubilizer are the main biofertilizers for horticultural crops. These micro-organisms are either free living in soil or symbiotic with plants and contribute directly or indirectly towards nitrogen and phosphorus nutrition of the plants. Motsara *et al.* (1995) reported that inoculation of *Azospirillum* and *Azotobacter* exerted beneficial effect on yield with varying physiological activities, including synthesis of plant growth promoting substances.

Biofertilizer is a cost effective renewable energy source and plays a crucial role in reducing the inorganic fertilizer application and at the same time increasing the crop yield besides maintaining soil fertility. In other words, biofertilizers are based on renewable energy sources and are eco-friendly compared to commercial fertilizers (Verma and Bhattacharyya, 1994). In the recent years, there is an urgent need to supplement the fossil fuel based inorganic fertilizers not only due to the hike in prices of chemical fertilizers but also a need is felt to maintain long term soil productivity and ecological sustainability.

Biofertilizer is a recent concept in horticultural crop. Generally fruit crops have now received more attention than vegetables and ornamental crops. *Glomus fasciculatum*, *Glomus mosseae*, *Azospirillum*, *Azotobacter* and phosphate solubilizing bacteria (PSB) are found useful for different horticultural crops. Use of biofertilizer particularly inoculation with *Azotobacter* could substitute 50% nitrogen requirement of banana and produce higher yield over full doses of nitrogen application (Tiwary *et al.*, 1999). The absorption of mobile nutrients like nitrogen also increases in association with VAM fungi (George *et al.*, 1992). Beneficial response of *Azotobacter* and *Azospirillum* in enhancing the productivity of banana was also reported by Dibut Alvorez (1996) and Mohandas (1996).

VAM fungi are responsible for more than two fold increased acquisition of the less mobile nutrient elements like P, Ca, S, Zn, Mg and Cu from the rhizosphere (Clark, 1997). The high efficiency of *Azospirillum* for fixing nitrogen and better mobilization of fixed phosphorus by VAM even at high temperature can make these highly suited for mosambi (Manjunath *et al.,* 1983). The per cent of wilting in VAM treated trees of guava was recorded to be lower as compared to untreated trees (Srivastava *et al.*, 2001). The root colonization per cent was high in *Glomus mosseae* inoculated papaya plants. Nutrient content of N, P, K and also of Fe, Mn, Zn and Cu increased due to VAM inoculation (Kennedy and Rangarajan, 2001).

The improvement in yield parameters in the presence of *Azospirillum* might be due to its dual nature in nitrogen fixation and production of phytohormones substances and increased uptake of nutrients such as nitrogen (Govindan and Purushothaman, 1984). Studies on biofertilizers along with chemical fertilizers were undertaken for assessing their effect on growth, yield and quality in Mosambi (Singh, 1990).

Effect of *Azotobacter* and *Azospirillum* on fruit crops

Mahmoud and Mahmoud (1999) reported improvement in plant vigour with inoculation of *Azospirillum* on peach seedling of cv. Nemaguard as compared to control. The treatment also led to increase in plant height, stem diameter, leaf number, plant dry weight and leaf area. Jeeva *et al.* (1988) in Tamil Nadu found that the inoculation of *Azospirillum* in combination with the nitrogenous fertilizer increased the yield upto 13.1% in Poovan. The increased bunch weight was also found associated with corresponding increase in length of bunch, number of hands, length, girth and weight of fingers.

Kerni and Gupta (1986) found greatest percentage increase in seedling height of mango, seedling diameter and number of leaves with treatment 49 g N, *Azotobacter* + 48 g N, 32 g N or *Azotobacter* alone as compared to control. Kumar and Shanmugavelu (1988) reported that both soil and foliar application of nitrogen and in combination with *Azotobacter* increase the plant height, plant girth, number of hands per bunch and number of finger/hands significantly in banana cv. Robusta.

The beneficial effect of *Azotobacter* inoculation in fruit and vegetable crops was well discussed by Asokan *et al.* (2000). Sharma (2002) in Assam revealed significant increase in the bunch weight and yield of banana with *Azotobacter* and organic manures supplements over 100% fertilizer.

Azotobacter also enhanced shooting and shortened crop duration. Wang in 1996 reported that with the application of *Azospirillum* + 150 kg N/ha can increase the yield in strawberry by 54%, the number of fruit per plant and clump weight were also highest compared to a treatment 150 kg N alone.

The microbial inoculants in combination with inorganic manures augmented the yield and nutrient uptakes in several crops (Chezhiyan *et al.*, 1999). They further reported increased bunch weight of 15.3 kg in hill banana var. 'Virupakshi' with application of biofertilizers (*Azospirillum*, Phosphobacteria and VAMF) and organic manure (FYM) along and with 75% NPK. Fernandez *et al.* (1998) reported that N-fixing bacteria improved pseudostem circumference and number of fingers/hand and advanced flowering time in banana. Mansour (1998) in Egypt found that apple trees treated with phosphorene active dry yeast and nitrobene at different concentrations found effective in improving fruit yield. The improvement was greatest with phosphorus biofertilizers. Similarly, Wang *et al.* (1998) reported that there is increase in number of fruits per plant, total weight of fruits and average fruit weight in strawberry as compared to the control by the application of *Azotobacter*, *Azospirillum* and P-solubilizing bacteria. Dalal *et al.* (2004) reported that the yield of the sapota is greatly increased due to the application of 75 kg FYM + 1500 g N + 1000 g Pps + 500 g kp + 12.5 g PSB. Benefit cost ratio is also high as compared to other fertilizer combinations.

Dibut Alvarez *et al.* (1996) in Cuba, studied the potential of *Azotobacter chrococcum* as a nitrogen fixer and bio-stimulant of banana and found that the bacteria inoculation along with N fertilizers between 80-100% favoured fruit development and also bacterial inoculation could compensate for 20% N fertilizer without changing the yield corresponding to 30 g N/ plantlets.

Vidhya (2004) found that application of 100 per cent of the recommended dose of fertilizer (RDF) along with 50 g in each of *Azospirillum*, phosphate solubilizing bacteria and Vesicular Arbuscular Mycorrhizae (VAM) in banana cv. Robusta resulted in better yield traits (Table 6.3).

TABLE 6.3

Influence of bio-fertilizers on yield of banana

Treatment	Bunch weight (kg)	No. of hands	No. of fingers	Finger weight (g)
100% RDF	24.47	9.14	137.08	164.85
100% RDF+ 50 g *Azospirillum*, PSB and VAM	26.10	10.63	142.75	173.36
SEd	0.34	0.21	0.94	2.06
CD (0.05)	0.75	0.46	2.04	4.49

Effect of Mycorrhiza on fruit crops

Singh and Singh (2004), reported that VAM significantly increase growth of plants compared to non-mycorrhizal control and was also effective in increasing nutrient uptake by the plants. VAM influenced growth attributing character and yield attributing component. About 50% saving of phosphorus was achieved through the use of VAM. Manjunath *et al.* (2001) reported that VAM fungi fasciculatum) was found to be effective in papaya in increasing the plant height, stem girth, petiole length and number of leaves. Rupnawar and Navale (2000) conducted an experiment on pomegranate and observed that mycorrhizal treatments were superior over non-mycorrhizal treatment in pomegranate. They reported that the *Glomus epigaeum* (GE) + *G. mosseae* + *Gigaspore calospora* mixture recorded the maximum height, root length, number of leaves, dry weight, of shoot and roots and mycorrhizal dependency percentage in pomogranate.

Mycorrhizal association with citrus species has been well established. Shamshiri and Usha (2004) found that application of *G. manihotis, G. gigaspora* and their combinations increased the trunk diameter, height and leaf area in Kinnow mandarin. Earlier, Sharma and Bhutani (1998) investigated on the response of VAM on apple seedlings in combination with VAM, *Azotobacter* and inorganic fertilizers. They reported that dual inoculation with *Glomus fasciculatum* and *Azotobacter chrococcum* produce larger plants which had a more leaf area. Kohli *et al.* (1998) reported that use of vermicompost, farm yard manure and biofertilizers like Azotobacter, Azospirillum, VAM increase production in citrus.

Effect of Biofertilizers on fruit quality

The use of biofertilizers found to influence the quality of the fruits. Singh *et al.* (2000) reported that the treatment combination of % P + VAM + N was the best treatment for producing better growth and yield of high quality fruit in Mosambi. This treatment also influence plant height, trunk diameter, canopy volume, root growth and biomass production as compared to control. In Egypt, the effect of biofertilizers on fruit set and productivity was investigated on Red Roomy grape vines (Akl *et al.*, 1997). The use of phosphorene was found to improve fruit set and yield as well as physical and chemical properties of fruits than control.

Suresh and Hasan (2001) in West Bengal evaluated the response of inoculation with Azospirillum and phosphobacteria on fruit quality of banana (Musa MA) cv. Giant Governor by manipulating the doses of nitrogen and potassic fertilizers. The results revealed that inoculation of biofertilizers along with the application of recommended dose of fertilizer proved most effective in improving fruit quality of Dwarf Cavendish banana cv. Giant Governor. Rana and Chandal (1999) reported that the plant growth, yield and fruit quality of strawberry were significantly increased with the application of biofertilizer and nitrogenous fertilizers. Maximum TSS content was observed with *Azotobacter* inoculation along with 80 kg N/ha.

Tiwary *et al.* (1998) found a fairly high TSS and reducing sugar content in fruits harvested from *Azotobacter* inoculated banana plant cv. 'GiantGovernor'. However, the effect of fertilizer in respect of total sugar and acidity content of fruit was not consistent. Sharma (2002) observed that application of *Azotobacter* + 75% inorganic N definitely improve the quality of banana *viz*., total sugar (16.88%), starch (2.28%) and protein (1.50%) and were recorded significantly higher the over recommended fertilizer dose. At lower pulney Hills, Saravanan (2007) investigated the effect of bioinoculants *viz., Azospirillum lipoferum,* phosphate solubilizing bacteria (*Bacillus megaterium*), *Arbuscular mycorrihizae* and *Pseudomonas fluorescens* in combination with farmyard manure and inorganic fertilizers on growth, yield and quality of mandarin orange during 2005-2007 and found 100 per cent RDF along with all bioinoculants registered higher fruit number, fruit yield with good quality fruits than the pure inorganically fertilized trees (Table 6.4).

TABLE 6.4

Effect of bioinoculants in combination with organic manure and inorganic fertilizers on yield and fruit quality of mandarin orange

Treatments	Fruit weight (g)	Number of fruits per tree	Yield per tree (kg)	TSS (Brix)	Ascorbic acid (mg/100 g)
Recommended dose of fertilizers (RDF) (600 : 200 : 400 g NPK plant^{-1})	91.50	100.97	9.00	8.89	21.80
100 per cent RDF + Bioinoculants*	84.50	147.95	12.90	9.44	23.13
75 per cent RDF + Bioinoculants*	86.00	140.95	12.00	9.51	22.47
CD (0.05)	1.75	8.06	0.08	0.26	0.62

* *Arbuscular mycorrhiza (500 g plant^{-1}) +Azospirillum lipoferum (100 g plant^{-1}) +Phosphate solubilizing bacteria (100 g plant^{-1}) + Pseudomonas fluorescens (100 g plant^{-1}).*

Integrated Nutrient Management

Experiment conducted in Tamil Nadu revealed that application of P enriched farm yard manure (0.5 kg P_2O_5 tree^{-1} mixed with 20 kg farm yard manure) along with 700 g N and 600 g K_2O tree^{-1} recorded the highest yield and improved the fruit quality of mandarin. Ingle *et al.* (2003) found that in Nagpur mandarin, number and weight of fruits and total soluble sugars were highest with application of 800 g N, 300 g P_2O_5, and 600 g K_2O along with 7.5 kg neem cake per plant per year. Tiwari *et al.* (1999) also obtained maximum yield of sweet orange with the application of 800 g N, 300 g P_2O_5, 600 g K_2O + 15 kg neem cake tree^{-1} year^{-1}.

Integrated nutrient management studies conducted at Andhra Pradesh revealed that 50% of the recommended dose of fertilizers besides 100g each of *Azospirillum* and Phosphobacteria + 5 kg of vermicompost resulted in higher fruit yield (15.48 t/ha) as against 13.98 t/ha with farmer's practice (Uma Jyothi *et al.*, 2004).

Sapota trees are fertilized with 100 kg farm yard manure, 6 kg castor cake and 2 kg super phosphate per tree (Narasimham, 1966); while in Maharashtra, ten years old trees are

TABLE 6.5

Effect of nitrogen through urea and castor cake on sapota

Treatment	Fruit weight (g)	Number of fruits per tree	Yield [kg tree^{-1}]	Fruit length [cm]	Fruit girth [cm]	Total sugars [%]	TSS [°Brix]
1200 g N from urea	59.05	2057.44	119.72	4.85	4.64	6.57	17.94
1200 g N (50% from castor cake + 50% from urea)	49.01	2453.56	131.24	4.71	4.57	6.68	18.22
1200 g N (75% from castor cake + 25% from urea)	47.56	2229.78	112.25	4.87	4.48	6.84	18.43
900 g N from urea	63.65	2627.56	163.30	5.07	4.75	6.69	18.48
900 g N (50% from castor cake + 50% from urea)	61.76	2485.67	151.12	5.21	4.80	7.06	18.59
900 g N (75% from castor cake + 25% from urea)	52.11	2385.44	124.46	4.95	4.73	7.10	18.40
750 g N from urea	53.30	2207.89	121.67	4.54	4.43	6.48	17.51
750 g N (50% from castor cake + 50% from urea)	42.97	2192.56	109.18	4.67	4.29	6.46	17.50
750 g N (75% from castor cake + 25% from urea)	52.21	2038.00	115.21	4.96	4.49	6.42	17.76
S. Em +	1.91	56.22	4.53	0.088	0.071	0.084	0.205
C. D. at 5%	5.42	160.35	12.90	0.25	0.20	0.24	0.58
C. V. %	10.68	7.35	10.66	5.39	4.66	3.77	3.40

fertilized with 40 kg farm yard manure and 6 kg of caster cake (Singh *et al.*, 1963). In Karnataka, the recommended dose for trees of 11 years and more is 40 kg farm yard manure, 400 g N, 160 g P and 450 g K per tree (Anonymous, 1975). According to existing knowledge, a sapota tree of 10 years and above age should be fed with 50 kg farm yard manure, 1000 g N, 500 g P_2O_5 and 500 g K_2O per annum.

Application of 5 kg vermicompost with 150 g N, 40 g P_2O_5 and 150 g K_2O per plant per year for 9 year old trees of Kalipatti at Karnataka and PKM-1 at Tamil Nadu conditions recorded significantly higher growth and yield, while at Gujarat conditions, application of 25 kg farm yard manure per 5 kg vermicompost with 300 g N, 50 g P_2O_5 and 200 g K_2O per plant for 15 years old Kalipatti sapota recorded significantly higher growth.

Inclusion of an organic source of nutrient (farmyard manure/neem cake) produced better quality fruits of papaya. Manivannan (2005) investigated the use of humic acid, an organic product derived from lignite on enhancing the biological growth, fruit yield and quality of papaya cv. Co-3. The study highlighted foliar spraying of humic acid at 0.2% concentration once in 15 days from the third month after planting improved the biometric characters, fruit yield and quality attributes *viz.* TSS, sugar acid ratio and ascorbic acid content besides enhancing the shelf life of the matured fruits.

In sapota, the highest fruit weight (63.65 g), the largest number of fruit per tree (2627.56) and the highest fruit yield (163.30 kg) were recorded in the combination in which 900 g N/tree was applied in the form of urea (T4). However, that was found on a par with the application of 900 g N/tree in which 50% N was from urea and 50% N from castor cake (T5). The highest reducing sugars and total sugars content (5.02 and 7.10%, respectively) were obtained in the treatment with 900 g N/tree in which 25% N was from urea and 75% N from castor cake (T6) (Table 6.6) (Devashi, 2012).

Apart from increasing yield, there is a need to produce high quality fruits to have larger market and preference by the consumers. Boora *et al.* (2002) reported increased total soluble solids and total sugar content with increased rates of inorganic fertilizer. Gawande *et al.* (1998) had reported the beneficial effects of organic manures when applied in conjunction with inorganic fertilizers in improving the quality of sapota fruits.

An experiment was conducted during 1998-2001 at Agricultural Research Station, Gangavati, Kamataka to study the effect of vermicompost application with varied levels of recommended dose of fertilizer on yield and quality of 'Kalipatti' sapota *[Manilkara zapota* (L.) P. Royan] and soil properties. The treatments were combinations of vermicompost @ 2.5 tonne/ha and 5 levels of recommended dose of fertilizer, *i.e.*, 100, 75, 50, 25 and 0% of recommended dose of fertilizer. The pooled data revealed that highest fruit yield was recorded under 100% recommended dose of fertilizer + vermicompost application (18.4 kg fruit/plant) that however remained on a par with 75% recommended dose of fertilizer + vermicompost and 100% recommended dose of fertilizer alone. The total soluble sugars remained significantly higher in all treatments involving vermicompost application (Hebbara *et al.*, 2006; Table 6.7).

TABLE 6.6
Effect of organic and chemical fertilizers on fruit set and yield

Treatment	No. of fruits/shoot	Yield/plant (kg)
Castor cake	46.25	81.25
50% C. cake + 50% RDF	48.75	84.00
FYM	54.50	85.00
50% FYM + 50% RDF	60.00	92.00
50% G. cake + 50% RDF	44.50	80.75
G. cake	44.50	78.50
RDF	50.75	86.00
CD at 5%	9.82	5.64

RDF—Recommended dose of fertilizer

TABLE 6.7
Influence of vermicompost and RDF levels on fruit quality of sapota

Treatment	TSS (Brix %)	Fruits/plant	Fruit weight (g)
100% RDF	21.7	313	50.6
100% RDF + VC	22.6	309	59.4
75% RDF + VC	22.3	294	55.0
50% RDF + VC	23.0	240	60.0
25% RDF + VC	22.8	203	50.2
VC only	22.9	181	54.0
CD (P = 0.05)	0.5	77.8	7.2

RDF—Recommended dose of fertilizer; VC—vermicompost

LEAF LITTER AS A SOURCE OF RECYCLE NUTRIENT

Plant litter, Leaf litter or tree Litter is dead plant material, such as leaves, bark, needles, and twigs that has fallen to the ground. Leaf litter is a rich resource of organic matter added to soil that decomposes to release nutrients are released to the environment. The portion of the litter that is not readily decomposable is known as humus. The animals, fungus and bacteria that live in and eat plant litter are called detritivores. Many studies have been conducted on leaf litter to evaluate the leaf litter quantity and nutrients content to determine energy fluxes and productivity. This is largely because leaf litter and its subsequent decomposition, strongly influences primary production and regulates energy flow and nutrient cycling in soil. Thus, leaf litter is a major participant in the transfer of energy and nutrients in a soil. Plant residues with different chemical composition show variable mineralization potential and decomposition behaviour (Mtambanengwe and Kirchman 1995).

Decomposition of leaf litter is catalyzed by microorganisms, which constitute part of the soil biomass. It acts as a source and sinks for the plant nutrients playing a crucial role in nutrient cycling and soil organic matter dynamics. It is the prime agent involved in plant residue decomposition, nutrient conservation and cycling processes in the soil (Singh *et al.*, 1989; Smith and Paul 1990). The microbial biomass has therefore been used as an index of soil fertility, which depends on nutrients fluxes (Hassink *et al.*, 1991). An increase in soil microbial biomass is considered essential for the improvement of soil fertility. Heterotrophic micro organisms act on organic materials and degrade them in soil. Consequently nutrients are made available in soil.

The decomposition of fine litter can be divided into three phases. The decomposition rate in the first phase is fast; last only few months, and is due to environmental conditions and litter quality. The second slow decomposition phases, last a few years and the third metastable phases last several decades (Trofymow *et al.*, 1998). Third phase theoretically begins following the attainment of a critical C/N ratio. The critical C/N ratio at which mineralization begins may vary with litter type and difference soil-litter systems (Berg and Staaf 1981; Hosseini and Azizi 2006). Biological activities *viz.* decomposition, mineralization etc. increase with increase in substrate, decomposable material and concentration. Similar results of increase in organic matter inputs have been reported by several workers (Singh and Gupta 1977, Swift *et al.*, 1979, Merekx and Mulonogy, 1994).

Even the trees that do not fix nitrogen can enhance physical, chemical and biological properties of soils by adding significant amount of above and belowground organic matter as well as releasing and recycling nutrients in tree bearing farmlands. The immobilized nutrients in microbial biomass constitute a reservoir of mineralizable N, and other nutrients, which upon biological transformation become available to plants (Singh *et al.*, 1989). However, the concentration and availability of nutrients under tree crown vary with depth of soil and distance from tree bole (Mazzarino *et al.*, 1991).

LEAF LITTER COMPOSITION OF FRUIT CROPS

Nutrients are released from decaying organic matter by physical (leaching) and biological (microbial decay) processes (Baker and Attiwill 1985). The sequence from fresh litter to stable soil organic matter has been defined as a decay continuum (Melillo *et al.*, 1989). Release/immobilization proceeds through time in different ways according to the mobility, concentration and biotic role of nutrients, the activity of organisms and the physical environment (Goya *et al.*, 2008).

The fertility of soil improves under the tree cover, which checks soil erosion, adds soil organic matter, available nutrients and replenishes the nutrients through effective recycling mechanisms.

Litter-fall is a major pathway for return of organic matter and nutrients from aerial parts of plant to the soil surface (Odiwe and Muoghalu, 2003). Litter production and decomposition rates have great importance in maintaining the fertility of soil. A substantial portion of nutrients accumulated by plants is returned to the soil as litter-fall followed by its decomposition. Standing crop of litter acts as an input–output system of nutrients and the rates at which litter falls and subsequently decays, regulate energy flow, primary productivity and nutrient cycling in forest ecosystems (Sundarpandian and Swamy, 1999; Pragasan and Parthasarathy, 2005).

NUTRIENT RELEASE PATTERN

The initial concentration of nutrients (N, P, K) differed significantly ($P < 0.01$) between leaf litters of different species. Initial nitrogen (N), phosphorous (P) and potassium (K) concentrations in alder leaf litter was significantly higher than the Khasi mandarin and Khasi pine leaf litter. The rate of release of nitrogen and phosphorus from decaying leaf litter of Alder occurred at much faster rate compared to Khasi pine while Khasi mandarin closely followed Alder. The release of nutrients from Pine litter occurred at the slowest pace. However, potassium release was faster from Khasi mandarin than other 2 species. The release of nitrogen from leaf litter ranged from 8.6 mg m^{-2} in Alder to 126.89 mg m^{-2} in Khasi pine. The stock of phosphorus in leaf litter decreased constantly with time and about 72, 81 and 34% of initial concentration was released from Alder, Khasi mandarin and Khasi pine leaf litter, respectively in 360 days. The percentage release of potassium from leaf litter ranged from 38% in Khasi pine to 69% in Alder. At the end of experiment 31 to 61% of the initial potassium stock of leaf litter remained undecomposed.

Several studies have reported that soil fertility under tree situation is improved due to increased input of organic matter through litter (Dunham, 1991; Kesseler, 1992; Campbell *et al.*, 1994; Dhyani, 1997). The leaf litter of Alder and Khasi mandarin with high concentration of N, P and K may therefore enhance fertility by adding high quality of litter to the soil system. The increase in soil total nitrogen, available phosphorus, potassium, water holding capacity, porosity and reduction in bulk densities of the topsoil are indicators of increased fertility under Alder based system. The Khasi mandarin based system was close to Alder based system in improving the soil fertility level. In Khasi pine based system, the poor quality of litter in terms of concentration of nutrients was compensated by greater amount of litter input.

TABLE 6.8

Canopy cover of different aged fruit plants

Tree species	Age (years)	Dbh (cm)	Avg. ht (cm)	Canopy cover (m^2)	Area under plantation (m^2)
Syzygium cumini	6	37.5-60.0	18	11.5-16.2	1800
Achras sapota	9	20.5-28.4	9	9.5-11.0	2400
Psidium guajava	9	17.6-22.5	9	8.0-12.5	3000

TABLE 6.9
Elemental composition of leaf of different fruit plants

Tree species	Total C (%)	Total N (%)	S (mg/kg)	P (%)	B (mg/kg)	Moisture %	C : N	C : S
P. guajava	51.0	0.7	0.5	0.7	0.7	2.5	30.0	94.17
A. sapota	52.2	0.4	7.2	0.9	0.9	4.5	36.5	7.26
T. indica	52.4	2.2	3.4	0.8	0.6	4.5	23.7	15.40
S. cumini	51.9	0.4	7.2	0.6	0.8	4.0	36.3	7.22

TABLE 6.10
Characteristics of the tree species (n = 5)

Tree characteristic	2007			2008		
	Mango	Avocado	Litchi	Mango	Avocado	Litchi
Height (m)	2.32	2.63	2.85	2.78	2.97	3.11
Stem diameter (cm)	9.40	11.8	11.8	10.8	13.2	13.4
Canopy width (m)	2.42	2.94	3.96	2.54	3.26	4.42
Litterfall (t ha^{-1})	3.52	4.22	8.04	7.61	8.43	8.46

Litter production varies according to habit of the tree species, its age and local environmental condition (Hawkins *et al.*, 1990; Szott and Kass, 1993; Table 6.8). Data on litter production indicates that major portion was contributed by leaf and several other workers reported similar findings from different forest stand (Singh, 1984; Sanchez and Sanchez, 1995; Pedersen and Hansen, 1999; Sundarpandian and Swamy, 1999; Zhou *et al.*, 2007).

The composition of leaf litter from different fruit species also differ widely and depends on soil properties and fertility status (Table 6.9). Characteristics of trees with respect to height and diameter also influences litter fall (Table 6.10). The litter standing crop represents the balance between input by litter fall and output by decomposition. Since climate of the study site was humid subtropical with distinct wet and dry seasons, litter fall and its decomposition was related to seasonal cycle. Leaf fall was confined to autumn and winter seasons, while decomposition was most rapid during wet period. The minimum litter mass on the ground during rainy season is therefore attributed to fast rate of decomposition owing to greater microbial activity due to favourable temperature and moisture conditions.

The rapid rate of decay after an initial lag phase is the net effect of a large number of processes such as utilization of readily available energy sources by microbes and loss of water-soluble components from leaf litter (Bloomfield *et al.*, 1993).

PLANT GROWTH PROMOTING RHIZOBACTERIA (PGPR)

The group of bacteria that colonize roots or rhizosphere soil and beneficial to crops are referred to as plant growth promoting rhizobacteria (PGPR).

The PGPR inoculants currently commercialized that seem to promote growth through at least one mechanism; suppression of plant disease (Bioprotectants), improved nutrient acquisition (Biofertilizers), or phytohormone production (Biostimulants). Species of *Pseudomonas* and *Bacillus* can produce as yet not well characterized phytohormones or growth regulators that cause crops to have greater amounts of fine roots which have the effect of increasing the absorptive surface of plant roots for uptake of water and nutrients. These PGPR are referred to as Biostimulants and the phytohormones they produce include Indole-acetic acid, cytokinins, gibberellins and inhibitors of ethylene production.

Recent advances in molecular techniques also are encouraging in that tools are becoming available to determine the mechanism by which crop performance is improved using PGPR and track survival and activity of PGPR organisms in soil and roots. The science of PGPR is at the stage where genetically modified PGPR can be produced. PGPR with antibiotic, phytohormone and siderophore production can be made.

Despite of promising results, biofertilizers has not got widespread application in agriculture mainly because of the variable response of plant species or genotypes to inoculation depending on the bacterial strain used. Differential rhizosphere effect of crops in harbouring a target strain or even the modulation of the bacterial nitrogen fixing and phosphate solubilizing capacity by specific root exudates may account for the observed differences. On the other hand, good competitive ability and high saprophytic competence are the major factors determining the success of a bacterial strain as an inoculant.

Studies to know the synergistic activities and persistence of specific microbial populations in complex environments, such as the rhizosphere, should be addressed in order to obtain efficient inoculants. In this regards, research efforts are made at Agricultural College and Research Institute, Madurai to obtain appropriate formulations of microbial inoculants incorporating nitrogen fixing, phosphate-and silicate-solubilizing bacteria and plant growth promoting rhizobacteria which will help in promoting the use of such beneficial bacteria in sustainable agriculture.

Recent studies confirmed that, a number of bacterial species mostly associated with the plant rhizosphere, are found to be beneficial for plant growth, yield and crop quality. They have been called 'Plant Growth Promoting Bacteria (PGPB)' including the strains in the genera Acinetobacter, Alcaligenes, Arthrobacter, Azospirillium, Azotobacter, Bacillus, Beijerinckia, Burkholdria, Enterobacter, Erwinia, Flavobacterium, Rhizobium and Serrotia (Bashan and de-Bashan, 2005).

The plant promoting effect of the PGPB is mostly explained by the release of metabolites directly stimulating growth. The mechanisms by which PGPB promote plant growth are not fully understood, but are thought to include : (*a*) the ability to produce plant hormones, such as auxins (Egamberdiyeva, 2005), cytokinins (Garcia de Salamone *et al.*, 2001), gibberellins (Gutierrez-Manero *et al.*, 2001), and inhibit ethylene production (Glick *et al.*, 1995); (*b*) asymbiotic N_2 fixation (Sahin *et al.*, 2004); (*c*) solubilization of inorganic phosphate and mineralization of organic phosphate and/or other nutrients (Jeon *et al.*, 2003); (*d*) antagonism against phytopathogenic microorganisms by production of siderophores, the synthesis of antibiotics, enzymes and/or fungidical compounds and competition with detrimental microorganisms (Lucy *et al.,* 2004).

Previous studies showed that PGPB stimulated growth and increased yield in several fruit species including apple, sweet cherry, citrus, raspberry, high bush blueberry, mulberry and apricot (Kloepper, 1994; De Silva *et al*., 2000; Sudhakar *et al.,* 2000; Esitken *et al*., 2002, 2003, 2006; Orhan *et al*., 2006; Aslantas *et al*., 2007; Karlidag *et al.,* 2007).

The effects of plant growth promoting bacteria (PGPB) on the fruit yield, growth and nutrient element content of strawberry cv. Fern were investigated under organic growing conditions between 2006 and 2008. The experimental plot was a completely randomized design with 3 replicates. Three PGPB strains (Pseudomonas BA-8, Bacillus OSU-142 and Bacillus M-3) were used alone or in combination as biofertilizer agent in the experiment. Data through 3 years showed that the use of PGPB significantly increased fruit yield, plant growth and leaf P and Zn contents. Root inoculation of M_3 and floral and foliar spraying of OSU-142 and BA-8 bacteria stimulated plant growth resulting in significant yield increases. M_3 + BA-8, BA-8 + OSU-142, M_3, M_3 + OSU-142 and BA-8 applications increased cumulative yield by 33.2%, 18.4%, 18.2%, 15.3% and 10.5%, respectively (Table 6.11). Number of fruits per plant significantly increased by the applications of M_3 + BA-8 (91.73) and M_3 (81.58) compared with the control (68.66). In addition, P and Zn contents of strawberry leaves with bacterial inoculation significantly increased under organic growing conditions (Esitken *et al.*, 2010).

TABLE 6.11

Yield, growth and fruit properties of strawberry cultivar when grown in 3 consecutive years under seven different PGPB application treatments on a calcareous Aridisol

	Control	BA-8	OSU 142	BA-8 + OSU 142	M_3	M_3 + BA-8	M_3 + OSU-142	M_3 + BA-8 + OSU-142
Cum. yield (g/plant)	481.9	532.6	518.5	570.6	569.5	642.1	555.4	490.1
Fruit weight (g)	7.11	7.06	7.13	6.99	6.98	6.98	6.85	6.66
Fruit firmness (kg/cm^2)	0.92	0.86	0.92	0.82	0.99	0.88	0.93	0.93
TSS(%)	9.03	8.66	8.99	9.08	7.74	8.62	8.65	8.27
Titr. acidity (%)	0.70	0.55	0.56	0.60	0.51	0.53	0.59	0.62
Vitamin C (mg/100 ml)	71.58	66.03	69.25	64.02	64.65	64.36	70.00	59.25
Reducing sugar (%)	5.51	5.15	6.02	6.49	5.18	5.39	5.84	5.44
Cum. fruit number	68.66	76.24	73.41	82.77	81.58	91.73	81.21	72.83
Leaf area (cm^2)	102.4	107.8	106.0	105.2	100.9	98.6	108.4	109.2

The experiments in this study showed that bacterial treatments significantly affected fruit yield, vitamin C and number of fruit per plant (Table 6.11). Significant yield increase was obtained with Bacillus M_3 (569.5 g/plant), Pseudomonas BA-8 (532.6 g/plant), BA-8 + OSU-142 (570.6 g/plant), M_3 + BA-8 (642.1 g/plant) and M_3 + OSU-142 (555.4 g/plant) treatments as compared with the control (481.9 g/plant) (Table 11). The percentage of yield increase was 33.2% when M_3 + BA-8 were applied.

MICROBIAL POPULATION IN RHIZOSPHERE SOIL OF FRUIT PLANTS IN COASTAL GUJARAT

A survey was conducted and soil samples were collected from rhizosphere of different fruit trees from coastal saline areas of Gujarat. The soil bacterial load from different fruit varied from 35×10^4 to 203×10^4. Various soil samples contain the large amount of microorganisms. From the CFU of the various soil samples, it was found that in Horticulture Plantation land use, the soil sample M8 contains the highest number of culturable microorganisms, because it shows approximately 203×10^4 CFU/g, while the soil sample of M_5 shows the lowest CFU/g (*i.e.* approximately 35×10^4 CFU/g). CFU of different soil samples varies from 35×10^4 CFU/g to 203×10^4 CFU/g (Table 6.12).

The characteristics of microorganism showed variation in size (large to small), shape (round or irregular) edge (undulate or entire), elevation (flat or convex) and gram reaction presence of gram +ve or gram –ve bacteria. No/Nil pigmentation reaction was observed. These native soil bacteria and fungi were found to have plant growth promoting characters that enhances plant growth and influences soil nutrient status on sustainable way.

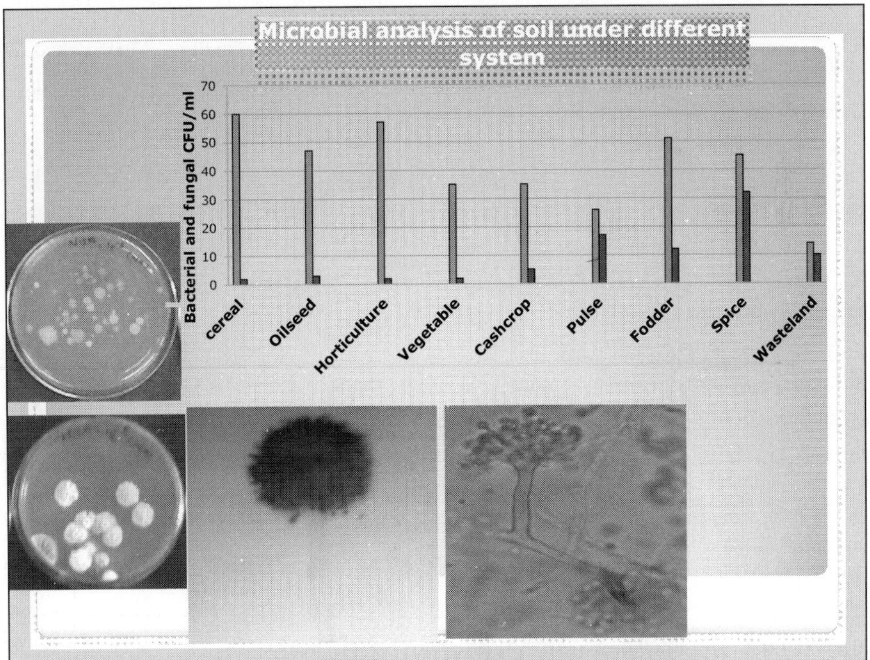

Fig. 6.3

TABLE 6.12
Bacterial load in rhizosphere soil

Sample orchard	CFU/g
Mango	173×10^4
S	35×10^4
M_8	203×10^4
M_{11}	76×10^4
M_{21}	39×10^4

REFERENCES

Akl, A.M. *et al.* (1997). Ann. Agric. Sci. **35** : 497-502.

Anandaraj, B. and Delapierre, L.R.A. (2010). Studies on influence of bioinoculants (*Pseudomonas fluorescens, Rhizobium* sp., *Bacillus megaterium*) in green gram. J. Biosci Tech, 1(2) : 95-99.

Anonymous (1975). *Cultivation practices of fruits*, UAS, Department of Horticulture, IIHR, Bangalore, pp. 31-34.

Aslantas, R., Cakmakci, R. and Sahin, F. (2007). Effect of plant growth promoting rhizobacteria on young apple tree growth and fruit yield under orchard conditions. Sci. Hortic. **111** : 371-377.

Asokan, R. *et al.* (2000). Indian Hort. **45** : 44-47.

Baker, T.G. and Attiwill, P.M. (1985). Loss of organic matter and elements from decomposing litter of *Eucalyptus obliqua* L'Hérit. and *Pinus radiata* D. Don. Australian Forest Research **15** : 309-319.

Berg, B. and Staaf, H. (1981). Leaching, accumulation and release of nitrogen in decomposing forest ?oor litter. In: Clark, F.E.; Rosswall, T. Terrestrial nitrogen cycles. Ecological Bulletin. **33** : 163-178.

Bloomfield, S., Williams, N., Lamb, D., Jackson, R. (1993). Non-weight bearing exercise may increase lumbar spine bone mineral density in healthy postmenopausal women. Am. J. Phys. Med. Rehab. **70** : 204-209.

Boora, R.S., Singh, Devi, Siddiqui, S. and Verma, S.L. (2002). Response of sapota to NPK fertilization. Haryana Journal of Horticulture Sciences **31**(1,2) : 15-17.

Chezhlyan, N. *et al.* (1999). South Indian Hort. **47** : 1-6.

Campbell, B.M., Frost, P., King, J.A., Mawanga, M., Mhlanga, L. (1994). The influence of trees on soil fertility on two contrasting semi-arid soil types at Matopos. Agro. Syst. **28** : 159-172.

Chowksey. (1991). M.Sc. (Ag.) thesis JNKVV. Jabalpur, India.

Clark, R.B. (1997). Plant Soil. **192** : 15-22.

Dalal, S.R. *et al.* (2004). PKVRes. J. **28** : 126-128.

Dibut Alvarez, B. *et al.* (1996). Infomusa. **5** : 20-23.

De Silva, A., Petterson, K., Rothrock, C. and Moore, J. (2000). Growth promotion of highbush blueberry by fungal and bacterial inoculants. Hort. Science **35** : 1228-1230.

Devashi, Varu (2012). Effects of organic and inorganic nitrogen on growth, yield and quality of sapota 'Kalipatti'. Journal of Fruit and Ornamental Plant Research **20**(1) : 55-64.

Dhyani, S.K. (1997). Analysis of the effect of agroforestry systems on productivity and soil characteristics. Ph.D. Thesis, North-Eastern Hill University, Shillong.

Dunham, K.M. (1991). Comparative effects of Acacia albizia and Kigelia africana trees on soil characteristics in Zambezi riverine woodlands. J. Trop. Ecol. **7** : 215-202.

Esitken, A., Karlidag, H., Ercisli, S., Turan, M. and Sahin, F. (2003). The effect of spraying a growth promoting bacterium on the yield, growth and nutrient element composition of leaves of apricot (*Prunus armeniaca* L. cv. Hacihaliloglu). Aust. J. Agric. Res. **54** : 377-380.

Esitken, A., Pirlak, L., Turan, M. and Sahin, F. (2006). Effects of floral and foliar application of plant growth promoting rhizobacteria (PGPR) on yield, growth and nutrition of sweet cherry. Sci. Hortic. **110** : 324-327.

Esitken, Ahmet, Hilal E. Yildiz, Sezai Ercisli, M. Figen Donmez, Metin Turan and Adem Gunes (2010). Effects of plant growth promoting bacteria (PGPB) on yield, growth and nutrient contents of organically grown strawberry. Scientia Horticulturae **124** : 62-66.

Fernandez, EM. *et al.* (1998). Fruits, **53** : 2.

Garcia de Salamone, I.E., Hynes, R.K. and Nelson, L.M. (2001). Cytokinin production by plant growth promoting rhizobacteria and selected mutants. Can. J. Microbiol. **47** : 404-411.

Gawande, S.S., Jiotode, D.J., Turkhede, A.B. and Darange, S.O. (1998). Effect of organic and inorganic fertilizers on yield and quality of sapota. Journal of Soils and Crops **8**(1) : 58-60.

Govindan, M. and Purushothaman, D. (1984). Agric. Res. J. Kerala. **22** : 133-138.

Goya, J.F., Frangi, J.L., Pérez, C., Tea, F.D. (2008). Decomposition and nutrient release from leaf litter in *Eucalyptus grandis* plantations on three different soils in Entre Ríos, Argentina. Bosque **29**(3) : 217-226.

Glick, B.R., Karaturovic, D.M. and Newell, P.C. (1995). A novel procedure for rapid isolation of plant growth promoting pseudomonas. Can. J. Microbiol. **41** : 533-536.

Gutierrez-Manero, F.J., Ramos-Solano, B., Probanza, A., Mehouachi, J., Tadeo, F.R. and Talon,M. (2001). The plant-growth-promoting rhizobacteria *Bacillus pumilus* and *Bacillus licheniformis* produce high amounts of physiologically active gibberellins. Physiol. Plant. **111** : 206-211.

Hassink, J. (1994). Active organic matter fractions and microbial biomass as predictors of N-mineralization. European Journal of Agronomy, **3** : 257-265.

Hassink, J., Lebbrink, G., Veen, V. (1991). Microbial biomass and activity of a reclaimed- polder soil under a conventional or reduced input farming system. Soil Biol. Biochem. **23** : 507-513.

Hawkins, R., Sembiring, H., Lubis, D. (1990). The potential of alley cropping in upland of east and central Java. UACP, Department of agriculture, Salatiga, Indonesia.

Hebbara, M., Ganiger, V.M., Masthana Reddy, B.G. and Joshi, V.R. (2006). Integrated nutrient management in sapota (*Manilkara zapota*) using vermicompost to increase yield and quality. Indian Journal of Agricultural Sciences **76** (10) : 587-590.

Hosseini, V. and Azizi, P. (2006). Determining rate of litter decomposition of Alnus subcordata in Asalem and Vaz regions by C/N index under laboratory conditions. J. Appl. Sci., **6**(1) : 40-42.

Ingle, H.V., Athawale, R.B. and Ingle, S.H. (2003). Effect of organic and inorganic fertilizer on yield and quality of Nagpur mandarin. Orissa J. Hort. **31**(1) : 10-13.

Jeeva, S. *et al.* (1988). *South Indian Hort.* **36** : 1-4.

Jeon, J.S., Lee, S.S., Kim, H.Y., Ahn, T.S. and Song, H.G. (2003). Plant growth promotion in soil by some inoculated microorganisms. J. Microbiol. **41** : 271-276.

Karlidag?, H., Esitken, A., Turan, M. and Sahin, F. (2007). Effects of root inoculation of plant growth promoting rhizobacteria (PGPR) on yield, growth and nutrient element contents of leaves of apple. Sci. Hortic. **114** : 16-20.

Kerni, P.N. and Gupta, A. (1986). Res. Develop. Reporter. **3** : 77-79.

Kennedy, Z.J. and Rangarajan (2001). Indian Phytopath. **54** : 72-77.

Kesseler, J.J. (1992). The influence of karate (*Vetellaria paradoza*) and nere (*Parkia biglobosa*) trees on sorghum production in Burkina Faso. Agro. Syst. **17** : 97-118.

Kloepper, J.W., 1994. Plant growth promoting bacteria (other systems). In : Okon, J. (Ed.), Azospirillum/Plant Association. CRC Press, Boca Raton, pp. 137-154.

Kohli, R.R. *et al.* (1998). Indian Hort. **43** : 12-15.

Kumar Ashok, A and Shanmugavelu, KG. (1988). Banana News. **11** : 11-12.

Lucy, M., Reed, E. and Glick, B.R. (2004). Application of free living plant-promoting rhizobacteria. Antonie van Leeuwenhoek **86** : 1-25.

Mahmoud, H.M. and Mohmand, EA.E (1999). Egyptian J. Hort. **26** : 7-18.

Manivannan,K. (2005). Influence of humic acid derived from lignite on fruit yield and quality in papaya(*Carica papaya* L.).*First International Symposium on Papaya*, Malaysia, Nov. 22-24, p. 76.

Manjunath, A. *et al.* (1983). Can. J. Bot. **61** : 2729-2732.

Manjunath, V.G. *et al.* (2001). J. Maharastra Agric. Univ. **26** : 269-271.

Mansour, AE.M. (1998). Egyptian J. Hort. **25** : 241-251.

Melillo, J.M., Aber, J.D., Linkins, A.E., Ricca, A., Fry, B. and Nadelhoffer, K.J. (1989). Carbon and nitrogen dynamics along the decay continuum: plant litter to soil organic matter. Plant Soil **115** : 189-198.

Merekx, R. and Mulonogy, K. (1994). Dynamics of organic matter in relation to the sustainability of agriculture systems. John Wiley and Sons, Chichester, UK.

Mohandas, S. (1996). In : Proc. Conference on Challenges for Banana Production and Utilization in 21st Century. National Research Centre on Banana. September 24-25, Trichy, Tamil Nadu, India, pp. 883-887.

Motsara, M.R. *et al.* (1995). In: Bio-fertilizer Technology Marketing and Uses-A Source Book cum Glossary. Fertilizer Development and Consultancy Organization, New Delhi, pp. 184.

Mtambanengwe, F. and Kirchmann, H. (1995). Litter from tropical savanna woodland (miombo) chemical composition and C and N mineralization. Soil Biol Biochem. **27** : 1639-1651.

Narasimham, (1966). Horticulture part I. Agricultural College, Bapatla, Andra Pradesh, pp. 200-201.

Narula, N. and Gupta, K.G. (1986). Ammonia excretion by *Azotobacter chroococcum* in liquid culture and soil in the presence of manganese and clay minerals. Plant and Soil **93** : 205-209.

Odiwe, A.I. and Muoghalu, J.I. (2003). Litterfall dynamics and forest floor litter as influenced by fire in a secondary lowland rain forest in Nigeria. Trop. Ecol. **44** : 241-248.

Okon, Y. and Labandera-Gonzalez, C. (1994). Agronomic applications of *Azospirillum*: an evaluation of 20 years worldwide field inoculation. Soil Biol. Biochem., **26** : 1591-1601.

Okon, Y. and Itzisohn, R. (1995). The development of *Azospirillum* as a commercial inoculant for improving crop yields. Biotechnol. Adv., **13** : 414-424.

Orhan, E., Esitken, A., Ercisli, S., Turan, M. and Sahin, F. (2006). Effects of plant growth promoting rhizobacteria (PGPR) on yield, growth and nutrient contents in organically growing raspberry. Sci. Hortic. **111** : 38-43.

Pedersen, L.B. and Hansen, J.B. (1999). A comparison of litterfall and element fluxes in even aged Norway spruce, sitka spruce and beech stands in Denmark. For. Ecol. Manage. **114** : 55–70.

Pragasan, L.A. and Parthasarathy, N. (2005). Litter production in tropical dry evergreen forests of south India in relation to season, plant life forms and physiognomic groups. Cur. Sci. **88** : 1255-1263.

Rana, R.K. and Chandal, J.S. (1999). Prog. Hort. **35**(1) : 25-30.

Rao, D.L.N. (1986). Nitrogen fixation in free living and associative symbiotic bacteria. In : Soil Microorganisms and plant growth. Subba Rao N.S. (Ed.) Oxford and IBH Pub. Co., New Delhi.

Roy, R.N. (1995). In: Integrated Plant Nutrition Systems. *FAO Fertilizer and Plant Nutrition Bulletin,* p. 3-22.

Rupnawar, B.S. and Navale, AM. (2000). J. Maharastra Agric. Univ. **25** : 44-46.

Sahin, F., Cakmakci, R. and Kantar, F. (2004). Sugar beet and barley yields in relation to inoculation with N_2-fixing and phosphate solubilizing bacteria. Plant Soil **265** : 123-129.

Sanchez, R.G. and Sanchez, J.A. (1995). Litter fall in primary and secondary tropical forests of Mexico. Trop. Ecol. **36** : 191-201.

Saravanan, T. (2007). Effect of Bioinoculants in combination with organic manure and inorganic fertilizers on growth, yield and quality of Mandarin orange (*Citrus reticulata* Blanco). M.Sc. Thesis submitted to TNAU, Coimbatore.

Sarig, S., Blum, A. and Okon, Y. (1988). Improvement of water status and yield of grown grain sorghum by inoculation with *A. brasiliense*. J. Agric. Sci. **110** : 271-278.

Shamshiri, M.H. and Usha, K. (2004). Effect of arbuscular mycorrhizal fungus on growth parameters in Kinnow. Improving Productivity, Quality, Post-harvest management and trade in horticultural crops. First Indian Horticulture Congress. pp : 243.

Sharma, R. (2002). Ph.D. Thesis, AAU, Jorhat.

Sharma, S.D. and Bhutani, I.P. (1998). Hort. J. **11** : 1-8.

Shen, J., Li, R., Zhang, F., Fan, J., Tang, C. and Rengel, Z. (2004). Crop yields, soil fertility and phosphorus fractions in response to long-term fertilization under rice monoculture system on a calcareous soil. Field Crop. Res. **86** : 225-238.

Singh, C. (1990). Ph.D. Thesis, IARI, New Delhi.

Singh, C. *et al.* (2000). Indian J. Hort. **57** : 114-117.

Singh, J. (1984). Effect of temperature, rainfall and soil moisture on soil respiration and litter decomposition in a subtropical humid forest ecosystem. Acta. Bot. India, **12** : 167-173.

Singh, J.S. and Gupta, S.R. (1977). Plant decomposition and soil respiration in terrestrial ecosystems. Botanical Rev. **43** : 449-528.

Singh, J.S., Raghubanshi A.S., Singh R.S. and Srivastava S.C. (1989). Microbial biomass acts as a source of plant nutrients in dry tropical forest and Savanna. Nature **338** : 499-500.

Singh, S., Krishnamurthy, S. and Katyal, S.L. (1963). Fruit culture in India, ICAR, New Delhi, pp. 92-98.

Smith, S.E. and Read, D. (1997). Mycorrhizal Symbiosis. Academic Press, San Diego.

Smith, J.L. and Paul, E.A. (1990). The significance of soil microbial biomass estimations. In: Bollag J-M, Stotzky G (eds) Soil biochemistry. Marcel Dekker, New York, pp 357-393.

Srivastava, A.K. *et al.* (2001). Indian Phytopath. **54** : 78-81.

Subba Rao, A. and Damodar Reddy, D. (2005). Meeting challenges of food security and soil quality. In: Perspectives of agricultural research and development. A compilation of articles for centenary seminar on recent advances in agricultural research held on 5.09.2005 at TNAU, Coimbatore.

Sudhakar, P., Chattopadhyay, G.N., Gangwar, S.K. and Ghosh, J.K. (2000). Effect of foliar application of *Azotobacter*, *Azospirillum* and *Beijerinckia* on leaf yield and quality of mulberry (*Morus alba*). J. Agric. Sci. **134** : 227-234.

Sundarpandian, S.M. and Swamy, P.S. (1999). Litter production and leaf-litter decomposition of selected tree species in tropical forests at Kodayar in the western Ghats, India. For. Ecol. Manage. **123** : 231-244.

Suresh, C.P. and Hasan, M.A. (2001). The Hort. J. **14** : 35-41.

Swift, M.J., Heal, O.W. and Anderson, J.M. (1979). Decomposition in terrestrial ecosystems. Blackwell Scientific, Oxford, UK, p. 372.

Szott, L.T. and Kass, D.C. (1993). Fertilizers in agroforestry. Agro. Syst. **23** : 157-176.

Tandon, H.L.S. (1997). In: Plant Nutrient Needs, Supply, Efficiency and Policy Issues : 2000-2005. NAAS, New Delhi, pp 15-28.

Tiwari, K.N. and Nagendra Rao, T. (2005). Changing face of soil fertility and nutrient management. In: Perspectives of agricultural research and development. A compilation of articles for centenary seminar on Recent advances in agricultural research held on 5.09.2005 at TNAU, Coimbatore.

Tiwary, D.K. *et al.* (1998). Agriculturist **42** : 235-240.

Tiwary, D.K. *et al.* (1999). Environ. and Ecology. **17** : 338-341.

Trofymow, J.A. (1998). The Canadian intersite decomposition experiment (CIDET): project and site establishment report, Canadian Forest Service, BC-X-378, pp. 126.

Uma Jyothi, K., Anka Rao, A. and Vishnu Sankara Rao, D. (2004). Integrated nutrient management strategies on yield and economics of sapota. ImproviOng Productivity, Quality, Post-harvest management and trade in horticultural crops First Indian Horticulture Congress. Pp. 257.

Verma, L.N. and Bhattacharyya, P. (1994). In : Production Distribution and Promotion of Biofertilizers, Fertilizers, Organic Manures. Recyclable Wastes and Biofertilizers. FDCo, New Delhi. pp. 132-147.

Vidhya, D. (2004). Studies on the effect of bio-fertilizers on banana cv. Robusta (AAA) and varying doses of inorganic fertilizers. M.Sc.(Thesis), Tamil Nadu Agricultural University, Coimbatore-3.

Wang, S.S. (1996). J. Soil Crops. **6** : 100-101.

Zhou, G., Guan, L., Wei, X., Zhang, D., Zhang, Q., Yan, J., Wen, D., Liu, J., Liu, S., Huang, Z., Kong, G., Mo, J., Yu, Q. (2007). Litterfall production along successional and altitudinal gradients of subtropical monsoon evergreen broadleaved forests in Guangdong, China. Plant Ecol. **188** : 77-89.

● ● ●

7

Biotechnological Interventions for Soil Health Improvement

JAVID IQBAL MIR, NAZEER AHMED, D.B. SINGH, ANIL SHARMA AND WAJIDA SHAFI

Biotechnology is defined as any technique that uses living organisms, or parts of organisms, to improve plants or animals, to develop microorganisms for specific uses, or to make or modify plant products. Crop productivity is related to the ability of plant roots to extract water and nutrients efficiently. The functioning of plant roots depends on an integrated set of biological processes. Biotechnological techniques can enhance the individual process at several points along this long pathway of biological steps. A very important, but often overlooked, aspect of applications of biotechnology to soil science problems is that it will require more detailed understanding of basic physical, chemical, and biological processes. Greater emphasis of study in those areas must proceed. Using this approach, we can identify areas of soil science where biotechnology will play a significant role in the future. Soil health generally covers all physical, chemical and biological parameters of soil, biotechnological interventions for improvement of these factors can improve the soil health which in turn effects the crop production. Biological component of soil can generally be exploited for biotechnological interventions. The role of biotechnology in soil microbiology and biochemistry includes genetic modifications of microorganisms for a host of applications. Some examples include the following :

Degradation of toxic organic substances in soil and water

Degradation of toxic organic substances can be enhanced by manipulating micro-organisms for improved activity of degradation enzymes. Degredation phenomenan can also be incorporated in plant roots and associated rhizosphere through genetic engineering tools. This phenomenon can decrease the level of toxic substances on one hand and can also increase the nutrient availability for plants on other hand.

Bioaccumulation of toxic trace elements from polluted soil/water

Plants are engineered for pumping of toxic trace elements into the plant system. Different genes have been utilized for channeling of toxic substances into the plants for cleaning of waste lands; this phenomenon is known as reclamation or soil remediation. Soil Remediation offers a state-of-the-art account of environmental biotechnology both in emerging and in more mature technological applications of soil remediation and cleanup of contaminated sites. Harnessing the potential of microorganisms and plants as eco-efficient and robust cleanup agents in a variety of practical situations is not only possible but is becoming widespread practice. Biotechnology plays an active role in bioremediation of contaminated soil, life cycle assessment software tools for remediation planning, *ex situ* cleanup technologies using slurry reactors, implementation of anaerobic and aerobic *in situ* processes including monitored natural attenuation, complementary technologies on pesticide immobilisation in soil or humification of nitroaromatics, and, finally, phytoremediation of recalcitrant organic compounds and heavy metals.

Waste management

Genetic engineering of plants and microorganisms improves the uptake, transport; partitioning, disposal and utilization of nutrients and toxic elements thus play an important role in waste management. Different genes have been incorporated into plant and microbial systems for increasing the uptake of different toxic substances and their further downstream channeling.

Development of stress tolerance in plants

Genetically engineered plants can develop tolerance to both biotic and abiotic stresses. This technology can help to develop plants which can thrive under different soil conditions. Plants have been engineered with genes like osmotin, LEA etc. which can help the plants to tolerate stresses like salinity, drought etc. Engineering of plants increases the chance of growing different crops under diverse soil conditions and make plants thrive well under different stress conditions.

RHIZOSPHERE DYNAMICS AND MYCORRHIZAE

Rhizobial dynamics and mycorrhizae association studies at molecular levels helps in identifying genes responsible for their association. Also their molecular diversity analysis helps in enhancing their utilization in crop improvement programs. The application of biotechnology in rhizosphere dynamics and mycorrhizae may be particularly important. Development will depend on close cooperation between microbiologists and plant nutritionists.

The future role of biotechnology lies not only in the genetic modification of plants. For biotechnology to be truly successful, we must strive toward a much better understanding of the molecular mechanisms of plant mineral nutrition and the genetic control of nutrient use efficiency. With this basic knowledge we can start thinking about genetically modifying plants for improved efficiency in uptake and utilization of nutrients, and for enhanced resistance to mineral stresses. The role of biotechnology in soil fertility and plant nutrition therefore involves the understanding of the molecular mechanisms of mineral nutrition of plants, genetic control of efficiency of nutrient use, genetic improvement of plants for (*i*) efficiency in uptake, transport, partitioning and utilization of nutrients; and (*ii*) resistance to mineral stresses; deficiencies, salinity, heavy metals, and toxic trace elements and rhizosphere dynamics and mycorrhizae.

Biotechnology will have a role in soil and water management and conservation. Applications will be important for molecular mechanisms of salt tolerance and of tolerance to heavy metals and toxic trace elements in plants, genetic improvement of resistance to salinity, soil acidity, heavy metals and to toxic trace elements.

Again, it is emphasized that although the focus is on genetic improvement of plants for resistance to adverse environmental conditions such as salinity, acidity, presence of heavy metals, or toxic trace elements, we should first understand the molecular mechanisms that can or cannot provide resistance to these mineral stresses. Therefore, we must focus on molecular mechanisms first, and then follow with research directed toward genetically improving plants. Biotechnology also will play a role in forest and range soils. The applications of biotechnology to forest soils may become a rapidly emerging field, specifically as it relates to the microbial component. The management of nutrient cycling in forest soils is very important. Progress will require a concerted effort by forest scientists, soil microbiologists, ecologists, and plant nutritionists. The possible genetic modification of major forest trees for improved resistance to acid deposition would mean tremendous progress in the quest for maintenance of forest ecosystems. Dinitrogen fixation in mixed legume-grass ranges has significance throughout the world.

Soil is one of the last great scientific frontiers (Science, 11 June, 2004) and the rhizosphere is the most active portion of that frontier in which biogeochemical processes influence a host of landscape and global scale processes. A better understanding of these processes is critical for maintaining the health of the planet and feeding the organisms that live on it (Morrissey *et al.*, 2004). There is a small but concerted effort under way to harness the root system of plants in an attempt to increase yield potentials of staple food crops in order to meet the projected doubling in global food demand in the next 50 years (Zhang, *et al.*, 2010; Gewvin, 2010). These efforts are being done in the face of a changing global climate and increasing global population which will inevitably require more productively grown food, feed and fiber on less optimal (and often infertile) lands; a condition already encountered in many developing countries (Tilman, *et al.*, 2002).

Biotechnological interventions can help in meeting the global challenges of food demand and nutritional security by improving the soil in following areas :

A. Rhizosphere

In 1904 the German agronomist and plant physiologist Lorenz Hiltner first coined the term "rhizosphere" to describe the plant-root interface, a word originating in part from the Greek word "rhiza", meaning root (Hiltner, 1904; Hartmann *et al.*, 2008). Hiltner described the rhizosphere as the area around a plant root that is inhabited by a unique population of microorganisms influenced, he postulated, by the chemicals released from plant roots. In the years since, the rhizosphere definition has been refined to include three zones which are defined based on their relative proximity to, and thus influence from, the root. The endorhizosphere includes portions of the cortex and endodermis in which microbes and cations can occupy the "free space" between cells (apoplastic space). The rhizoplane is the medial zone directly adjacent to the root including the root epidermis and mucilage. The outermost zone is the ectorhizosphere which extends from the rhizoplane out into the bulk soil. Newman (1985) examined a variety of plant species and estimated that roots can release anywhere from 10 to 250 mg C/g root produced or about 10-40% of their total photo synthetically fixed carbon. The C released is in both organic (*e.g.*, low molecular weight organic acids) and inorganic (*e.g.*, HCO_3) forms, however, the organic forms are the most varied and can have the most influence on the chemical, physical and biological processes in the rhizosphere (Jones, *et al.*, 2009). The composition and amount of the released compounds is influenced by many factors including plant type, climactic conditions, insect herbivory, nutrient deficiency or toxicity, and the chemical, physical and biological properties of the surrounding soil. The root products imparted to the surrounding soil are generally called rhizodeposits. Rhizodeposits have been classified based on their chemical composition, mode of release, or function but are classically defined (Rovira, 1969) to include sloughed-off root cap and border cells, mucilage, and exudates. Components in root exudates assist plants in accessing nutrients by acidifying or changing the redox conditions within the rhizosphere or directly chelating with the nutrient. Exudates can liberate nutrients via dissolution of insoluble mineral phases or desorption from clay minerals or organic matter where they are released into soil solution and can then be taken up by the plant.

Biotechnology can play an important role in modifying the plants by genetic engineering to secrete specific root exudates which can help in acquiring nutrients (Fe, P etc.), generates invasiveness (allelopathy), generate chemical signals to attract symbiotic partners (chemotaxis), to decide microbial composition of rhizosphere etc. Genetic engineering of plants for manipulating root exudates can also play an important role in nutritional solubolize the nutrients by displaying some bioactive compounds like siderophores. The phytosiderophore chelates strongly with Fe and is then brought back to the root via diffusion where plasma membrane transporters specific to the chelated Fe shuttle it into the cells.

B. Microbe-Plant Relationships

Rhizodeposits make the rhizosphere a desirable niche for microbial communities to proliferate. One teaspoon of bare or tilled soil contains more microorganisms than there are people on Earth, however, the rhizosphere can have 1000-2000 times that number

(10^{10}-10^{12} cells per gram rhizosphere soil) making it a pretty crowded place. Nutrient availability for microbial growth is higher in the rhizosphere compared to the bulk soil; however, many microbes are competing for these nutrients, some more successfully than others. The plant rhizosphere-microbe relationships that have received the most attention include those of Rhizobia bacteria and their symbiotic plant partners, mychorrhizal fungi associations, and beneficial plant growth promoting rhizobacteria (PGPR).

B.1. *Legume-Rhizobia Symbiosis*

The observation that including leguminous plants in crop rotations results in better yields of non-legumes goes back many centuries to the Romans and Greeks. It was not until 1888, however, that Hellriegel and Wilfarth (1888) definitively proved that the cause of the improved yield was from the conversion of atmospheric dinitrogen ($N_{2(g)}$) into ammonia, a form usable by the plant, by rhizobium bacteria (*Rhizobium leguminosarum*) housed within nodules on the roots of leguminous plants (Hirsch, *et al.*, 2001). This discovery inspired Lorenz Hiltner who spent his career researching and developing rhizobial inoculants to improve agricultural productivity eventually leading to his development of the Rhizosphere concept. The interaction between rhizobia and legumes has been exhaustively studied since that time, elucidating several facets of the interaction.

The genes involved in plant-microbe interaction, their activation, expression, regulation and role needs to be studied completely. Both genes of plant (legume) and microbe have their roles in nutrient fixation, here biotechnology and molecular biology plays an important role in deciphering the pathways affected during the process. Hence the key genes *e.g.* Nod genes, Nif genes etc. can be used for genetic engineering of these organisms for enhancing nutrient fixation efficiency. During nitrogen fixation main barrier for association is pectin of plant roots which hampers the process of nodulation but genes encoding pectate lyase (Pel and Peh genes) if transferred to these legumes can improve process of nodulation by enhancing pectin degradation. Nodulation process is also influenced by different signals like flavonoid signal which induces nodulation genes (nod genes) in the rhizobia which encode enzymes necessary to produce a chemical response (lipochitooligosaccharide) to the plant known as a nod factor. The nod factors initiate a cascade of developmental processes in the plant root which allow for the invasion of the bacteria and formation of the nodule in which the bacteria are eventually housed (Jones *et al.*, 2007). The structure of the nod factors are largely species specific and are one of the reasons for the host specificity observed between rhizobia and their plant partners (Giles, *et al.*, 2008). This specificity is quite astonishing considering that one gram of soil (about the size of a raisin) contains > 10^9 bacteria from 10,000 different species, from which a miniscule amount have developed the ability to symbiotically cohabitate with plants.

B.2. *Mycorrhizal Fungi and Nutrient Acquisition*

Mycorrhiza (from the Greek words for fungus and root) is a general term describing a symbiotic relationship between a soil fungus and plant root. Unlike rhizobia and their legume partners, mycorrhizal associations are ubiquitous and relatively non selective, occurring in

~ 80% of angiosperms and in all gymnosperms (Wilcox, 1991). The ability for plants to form symbiotic associations with mycorrhizae occurred early in the evolution of plants (~450 million years ago) compared to legumes (~ 60 million years ago), which is likely why mycorrhizae are now ubiquitous throughout the plant kingdom. Although parasitic and neutral relations exist, a majority of these associations are beneficial both to the host plant and the colonizing fungi. Mycorrhiza assist plants in obtaining water, phosphorus and other micronutrients (*e.g.*, Zn and Cu) from the soil and in return receive sustenance (carbon) from the plant.

There are two broad categories of mycorrhizal associations with plant roots,

- Ectomycorrhiza
- Endomycorrhiza

B.2.1. *Ectomycorrhizae*

The ectomycorrhizae (EM) occur mainly in the roots of woody plants (*i.e.* forest trees) and form a dense hyphal covering (fungal sheath or mantel) over the root tip from which hyphae grow into the intercellular spaces forming a net (Hartig net) of hyphae around the root cortex cells, but do not penetrate the cell walls.

B.2.2. *Endomycorrhiza*

The endomycorrhizae fungal hyphae grow into the root cortex and enter the cells forming fan-like, highly branched structure known as an arbuscule that remain separated from the cytoplasm by the plant plasma membrane (Harrison, 2005). The endomycorrhiza can be further divided into the more widespread arbuscular mycorrhiza (AM) and the specialized orchid and ericoid mycorrhizas which, as the name implies, are colonizers of orchids and ericoid (*e.g.*, cranberry) plant species. The AM fungal associations are the most abundant of all mycorrhizal associations.

In both cases, the Hartig net and the arbuscules increase the contact area between the fungus and the plant through which the transfer of nutrients to the plant and carbon to the fungus occurs. Unlike the ectomycorrhiza, the endomycorrhiza are wholly dependent on the plant for their carbon and when associations occur, both endomycorrhiza and ectomycorrhiza can demand up to 20-40% of the total photosynthetically fixed carbon the plant produces.

The chemical dialogue that initiates mycorrhizal associations with plants is intricate and not as well understood as that of the Rhizobia. In part, the lack of understanding rests in the fact that endomycorrhiza are obligate symbionts, meaning they cannot be grown independent of their plant hosts, which makes it difficult for scientists to study them in the lab under controlled conditions. For this reason, more is known about the development of the ectomycorrhizal symbiosis because the two partners can be grown independently. In both cases, there is evidence that the fungus can sense the presence of volatiles (CO_2), or chemicals in plant root exudates which then initiates hyphal growth and branching (Martin *et al.*, 2001). Conversely, chemical signals from mycorrhizea which are perceived by the plant have long been suspected (Kosuta *et al.*, 2003) and only recently has the structure of one of these chemicals in the ecotmycorrhizal association been fully characterized (Maillet *et al.*, 2011). The chemical signals are called *Myc*-factors which, interestingly, are similar in structure to the nod factors produced by rhizobia in the rhizobia-legume symbiosis. It is still unclear whether the fungus produces the signals all the time or only when it senses the plant is near. Regardless, fungal spores in the soil remain in a resting state until one of these stimuli initiates germination and hyphal growth. If close to a root, hyphal growth and branching will increase and *Myc*-factors will initiate root branching presumably to improve the odds of root-fungus interception. Identification and role of different genes and pathways involved during the interaction will play an important role in enhancing the utilization of this association for improvement of different crops.

Genetic diversity of different mycorrhizal strains using different molecular markers will be helpful in identifying species specific mycorrhizal strains. Genetic engineering of plants can also widen their association circle with different strains of mycorrhizae. Exudate composition responsible for successful mycorrhizal association can be modified through biotechnological tools to increase the number of partners for association.

Biotechnology utilizes the strength of molecular markers as diagnostic tools for identification of specific strains at any developmental stage. Strain specific molecular markers will help in screening/identification of different mycorrhyzal strains suitable for specific hosts at their earlier stages of development; and hence can be utilized for further association and crop improvement.

C. Plant Growth Promoting Rhizobacteria (PGPR)

Plant growth promoting rhizobacteria were first defined by Kloepper and Schroth (1978) as organisms that, after being inoculated on seeds, could successfully colonize plant roots and positively enhance plant growth. Till date, there have been over two dozen genera of nonpathogenic rhizobacteria identified. Plant growth promotion can be shown to work directly on the plant in the absence of root pathogens by the release of plant growth stimulating compounds (*e.g.* phytohormones such as auxins or cytokinins) and improvement in mineral uptake (*e.g.* siderophore release increasing Fe availability). Plant growth promotion can also occur indirectly by control of pathogens (biocontrol) via synthesis of antibiotics or secondary metabolite-mediated induced systemic resistance (ISR) (van Loon, *et al.*, 1998, 2007).

Microbial colonization of the plant root surface is not uniform, but instead occurs in patches along the root, ultimately covering ~15-40% of the total plant root surface. The density and structure of the microbes on the root surface are dictated by nutrient availability and physicochemical variations throughout the root surface. Root exudates can serve as a food source and chemoattractant for microbes which then attach to the root surface and form microcolonies. Common sites for bacterial attachment and colonization are at epidermal cell junctions, root hairs, axial groves, cap cells, and sites of emerging lateral roots (Danhorn *et al.,* 2007). Microcolonies can eventually grow into larger biofilms consisting of multiple layers of bacteria which are encased in an exopolymeric matrix. In many cases the effectiveness of rhizobacteria at promoting growth occurs in a density dependant manner (Rudrappa, *et al.*,

2008). Once a critical microbial density is reached, the biofilms then begin to act in unison, in a process known as quorum sensing, coordinating the release of compounds that aid in the promotion of plant growth via the direct and indirect mechanism. Biotechnological interventions can boost production of plant growth stimulating compounds on one hand and can also enhance the colonization of microbial population around the rhizosphere. Genetically modified plant growth promoting rhizobia with improved growth promotion phenomenon have successfully been developed the genetic engineering of these microbes.

D. Root System Architecture—Sensing and Foraging

The spatial extent and influence of the rhizosphere are dictated by root system architecture (RSA). Root system architecture is very malleable, determined by plant species and occurring in response to changing climactic, biological and edaphic conditions. The distribution of nutrients in soils is heterogeneous or patchy and there is evidence plants can "sense" the presence of nutrients and allocate more resources to the root system and direct root growth toward or within these patches. Examples include increased root depth in drought tolerant beans, wheat, and maize and an increase of dense shallow root systems leading to topsoil *foraging* for P in plants under low P conditions (Ho *et al.*, 2005). The nutrients themselves and the activity of specific molecular pathways involved in their acquisition may act to induce hormonal signals triggering increases in root density and length (Lopez-Bucio *et al.*, 2003). Because of the underground nature of plant roots, uncovering the mechanisms responsible for root foraging has been difficult, but full annotation of the *Arabidopsis* genome and a complete map of root cell types have dramatically increased our understanding of root biology and these processes (Malamy and Benfey, 1997; Dinneny, *et al.*, 2008; Benfey *et al.*, 2010). Genes responsible for root system architechture have been identified, their complete role and regulation have been studied. Thus these genes can be utilized for development of need based root system architecture to maximize nutrient uptake and crop production.

E. Soil Microbial Communities and Metagenomics

Metagenomics Coined by Jo Handelsman and others in the University of Wisconsin, Department of Plant Pathology in 1982 (Handelsman, 2004), metagenomics is a new field of research that allows the study of genomes recovered from environmental samples. Metagenomics can be defined as "the application of modern genomics techniques to the study of communities of microbial organisms directly in their natural environments, bypassing the need for isolation and laboratory cultivation of individual species (Chen and Pachter, 2005). Metagenomics is the culture independent analysis of a mixture of microbial genomes (Schloss and Handelsman., 2003; Riesenfeld *et al.,* 2004; Susannah and Edward., 2005; Patrick and Handelsman., 2005) The metagenomic methodology has been developed as an effective tool for the discovery of new natural products and microbial functions (He *et al.*, 2007) it can be described as the application of the genomics suit of technologies to uncultivated microorganisms (Gabor *et al.,* 2007; Singh *et al.*, 2008). Studies have revealed that only 0.001-0.1% of the total microbes in sea water, 0.25% in freshwater, 0.25% in sediments and only 0.3% of soil microorganisms could be cultivable in vitro (Amann *et al.,* 1995; Singh *et al.*, 2008). The current metagenomic studies have largely progressed due to the construction of efficient gene cloning vectors like bacterial artificial chromosomes (BACs) or cosmids, (Xu, 2006; Babcock *et al.*, 2007) which allow cloning and expression of larger and complex DNA segments or genes and the development of methods for generation and analysis of the data (Singh *et al.*, 2008; Ghazanfar and Azim, 2009).

Soil is a complex environment, which is a major reservoir of microbial genetic diversity (Robe *et al.*, 2003). Soil is dominated by the solid phase (Rolf, 2004) and the soil microorganisms are localized in close association with soil particles, *e.g.* with complexes of clay–organic matter (Foster, 1998). The microorganisms can be found as single cells or microcolonies. Their metabolism and interactions with other organisms and with soil particles is dependent on the conditions at the microhabitat level, which often differ between microhabitats even over very small distances. The microhabitats for soil microorganisms include micropores and the surfaces of soil aggregates of various composition and sizes (Ranjard and Richaume, 2001; Torsvik *et al.*, 2002). Thus, soil can be regarded as very heterogeneous with respect to conditions for microbial growth and for the distribution of microorganisms and matrix substances. This heterogeneity results in a wide variety of microbial niches and a high diversity of soil microorganisms. The microbial diversity in soils exceeds that of other environments and is far greater than that of eukaryotic organisms. One gram of soil can contain up to 10 billion microorganisms of possibly thousands of different species (Rosello and Amann, 2001). The genetic complexity of microbial soil communities has been estimated by re-association of community DNA. Not taking into account the genomes of rare and unrecovered microorganisms, such analyses have shown that the soil community size is equivalent to 6,000-10,000 *Escherichia coli* genomes (Ovreas, 2000; Torsvik *et al.*, 2002). While analysing the re-association kinetics of the total bacterial DNA in a 30 g soil sample Doolittle (1999) reported that it contained more than 500 000 species. These results illustrate that the genetic diversity of the soil metagenome is a rich and widely unexplored resource for new industrial enzymes and bioactive compounds. Simple calculations of soil microbial diversity place it in the range of between 3,000 and 11,000 genomes per gram of soil with less than 1% being accessible through cultivation techniques (Torsvik and Ovreas, 2002; Torsvik *et al.*, 2002; Curtis and Sloan, 2004).Soil metagenomics can provide a cultivation-independent assessment of the largely untapped genetic reservoir of soil microbial communities. Culturable microorganisms provide very limited information on soil microbial diversity, because only 0.1-10% of soil microorganisms can be cultured under conventional laboratory conditions. Soil metagenomic methods, comprising isolation of soil DNA and construction and screening of clone libraries, enable researchers to peep at more complete scenario of soil microbial communities, and thus, to better understand their interactions.

Application of soil metagenomics

The development and application of metagenomics has enabled access to the uncultivated soil microbial community, availing a rich source of novel and useful biomolecules. Some examples of application of soil metagenomics are :

- **Antibiotics and pharmaceuticals.** Metagnomic study will help in identifying the clones responsible for synthesis of different antibiotics or other bio-active compounds. The genes responsible for the synthesis of specific antibiotic or other compounds can be identified from specific microbe present in the soil sample used for matagenomic analysis through biotechnological tools.
- **Enzymes.** Enzymes encoding genes can be screened from metagenomic libraries and can be further utilized for downstream application.
- **Vitamins.** Identification of novel genes for biosynthesis of vitamins has been done through metagenomic approaches.
- **Crop specific rhizosphere.** Metagenomic studies are used for identifying genetic diversity of microbes associated with rhizosphere of different crops. Genetic diversity studies under different growing conditions and among different varieties of a single crop are also done.
- **Synthesis of bioactive compounds.** Metagenomic studies help in identifying genes responsible for biosynthesis of different bio-active compounds like plant growth promoters, inhibitors, hormones etc.

F. Biofilms

Biofilms in soil are composed of multiple species microbial consortia attached to soil particles and biotic surfaces including roots, fungal hyphae and decomposing organic material. The bacteria present in these biofilms gain several advantages including protection from predation, desiccation and exposure to antibacterial substances, and optimized acquisition of nutrients released in the mycosphere. Studies of soil biofilms are complicated by the composite structure of the soil environment; therefore, various simplified model systems have been applied to study succession and bacterial interactions in soil biofilms. Model system observations indicate an increased efficiency to degrade and decompose organic material and xenobiotic compounds by these multispecies bacterial communities. Consequently, soil biofilms may be valuable tools for bioremediation and biocontrol. However, soil biofilms may also provide survival sites for opportunistic pathogenic bacteria, providing enhanced protection and increasing their potential to survive and evolve in the soil environment. Hence biofilms are of major importance for the fitness of individual bacteria and the wider soil ecology, due to the accumulated selective advantage provided to bacteria by the biofilm mode-of-life.

Biofilm system is continuously drawing great attention for research. Based on the reports, it may be concluded that biofilms have the potentiality to remove heavy metals and other xenobiotic compounds from wastewaters and natural waters containing low levels of pollutants. Moreover, the tendency to concentrate the pollutants in biofilm in natural environment can serve the negative function of providing a mechanism of entry of pollutants into the aquatic organism food chain. This will provide an opportunity to reflect on the role of biofilm towards wastewater treatment and environmental research. Further research area needs to be extended on the focus of gene transfer within biofilms. Study of biofilm communities and gene transfer within biofilms would facilitate the development of better techniques for the bioremediation of polluted sites and wastewaters.

G. Biosensors–highly sensitive and effective detectors

Biotechnology not only plays an important role in remedying soil damage, but also in the detection of soil damage. The successful cooperation between actors in the biotechnology and microsystems technology sectors has led to huge progress in biosensor technology over the last few years. Biosensors consist of a biological recognition element in the majority of cases a biochemical receptor and a transducer that converts the receptor's recognition reaction into an electronically usable signal. The biosensors' strengths are their outstanding sensitivity and selectivity. They can help find the virtual needle in a haystack. For example, biosensors can be used to determine the bacteria concentration in freshwater used for bathing and they can also detect even the smallest concentrations of arsenic.

REFERENCES

Morrisseey, J., Dow, J., Mark, G., O'Gara, F. (2004). Are microbes at the root of a solution to world food production? *EMBO rept.* **5**(10) : 922-926.

Zhang, F., Shen, J., Zhang, J., Zuo, Y. Li, L., Chen, X. (2010). Rhizosphere processes and management for improving nutrient use efficiency and crop productivity: implications for China. Advances in Agronomy. **107** : 1-32.

Gewin, V. (2010). Food : An underground revolution. Nature **466** : 552-553.

Tilman, D., Cassman, K.G., Matson, P.A., Naylor, R. and Polasky, S. (2002). Agricultural sustainability and intensive production practices. Nature. **418** : 671-677.

Hiltner, L. (1904). Ueber neuere Erfahrungen und Probleme auf dem Gebiete der Bodenbakteriologie und unter besonderer BerUcksichtigung der Grundungung und Brache. Arb. Deut. Landw. Gesell, **98** : 59-78.

Hartmann A., Rothballer M., and M. Schmid. (2008). Lorenz Hiltner, a pioneer in rhizosphere microbial ecology and soil bacteriology research. Plant Soil **312** : 7-14.

Newman, E.I. (1985). The rhizosphere : carbon sources and microbial populations. Ecological Interactions in Soil. A. H. Fitter. Oxford, Blackwell Scientific Publications : p. 107.

Jones, D.L., Nguyen, C., and Finlay, R.D. (2009). Carbon flow in the rhizosphere : carbon trading at the soil-root interface. Plant Soil. **321** : 5-33.

Rovira, A.D. (1969). Plant root exudates. Botanical Review. **35**(1) : 35-57.

Hellriegel H, Wilfarth H. (1888). Untersuchungen uber die Stickstoffnahrung der Gramineon und Leguminosen. Beilageheft zu der Ztschr. Ver. Ru¨ benzucker-Industrie Deutschen Reichs.

Hirsch, A.M., Lum, M.R. and Downie, J.A. (2001). What makes the rhizobia-legume symbiosis so special ? Plant. Phys. **127** : 1484-1492.

Jones, K.M., Kobayashi, J., Davies, B.W., Taga, M.E., and Walker, G.C. (2007). How rhizobial symbionts invade plants: the Sinorhizobium-Medicago model. Nature Reviews Microbiology. **5**(8) : 619-633.

Giles E.D. Oldroyd and J. Allan Downie (2008). Coordinating Nodule Morphogenesis with Rhizobial Infection in Legumes. Annual Review of Plant Biology. **59** : 519-546.

Wilcox H.E. (1991). Mycorrhizae. In : Waisel Y, Eshel A, Kafkaki U, eds. Plant roots : the hidden half. New York, USA : Marcel Dekker.

Harrison M.J. (2005). Signaling in the arbuscular mycorrhizal symbiosis. Annu. Rev. Microbiol. **59** : 19-24.

Martin, F., Duplessis, S., Ditengou, F., Lagrange, H., Voiblet, C., and Lapeyrie, F. (2001). Developmental cross talking in the ectomycorrhizal symbiosis: signals and communication genes. New Phytologist. **151** : 145-154.

Kosuta, S., Chabaud, M., Lougnon, G., Gough, C., Denarie, J., Barker, D.G., Becard, G.A. (2003). diffusible factor from arbuscular mycorrhizal fungi induces symbiosis-specific MtENOED11 expression in roots of Medicago truncatula. Plant Physiol. **131** : 952-962.

Maillet, F., Poinsot, V., Andre, O., Puech-Pages, V., Haouy, A., Gueunier, M., Cromer, L., Giraudet, D., Formey, D., Niebel, A., Andres Martinez, E., Driguez, H., Cecard, G., Denarie, J. (2011). Fungal lipochitooligosaccharide symbiotic signals in arbuscular mycorrhizal. Nature. **469** : 58-64.

Kloepper, J.W. and Schroth, M.N. in Proc 4th int. Conf. Plant Pathogenic Bacteria Vol. 2 (ed. Station de Pathologie Végétale et Phytobactériologie) 879-882 (Gibert-Clarey, Tours, 1978)

Van Loon, L.C. (2007). Plant responses to plant growth-promoting rhizobacteria. Eur. J. Plant Pathol. **119** : 243-254.

Van Loon, L.C., Bakker, P.A.H.M. and Pieterse, C.M.J. (1998). SYSTEMIC RESISTANCE INDUCED BY RHIZOSPHERE BACTERIA. Annual Review of Phytopathology. **36** : 453-483.

Danhorn T, Fuqua C. (2007). Biofilm formation by plant-associated bacteria. ANNUAL REVIEW OF MICROBIOLOGY. **61** : 401-422.

Rudrappa, T., Biedrzycki, M.L. and Bais, H.P. (2008). Causes and consequences of plant-associated biofilms. FEMS Mirobial. Ecol. **64** : 153-166.

Ho, M.D., Rosas, J.C., Brown, K.M., Lynch, J.P. (2005). Root architecture tradeoffs from water and phosphorus acquisition. Functional Plant Biology. **32** : 737-748.

Lopez-Bucio J, Cruz-Ramirez A, Herrera-Estrella L. (2003). The role of nutrient availability in regulating root architecture. Current Opinion in Plant Biology **6**(3) : 280-287.

Malamy, J.E. and Benfey, P.N. (1997). Organization and cell differentiation in laterl roots of Arabidopsis thaliana. Development. **124** : 33-44.

Dinneny J.R., Long T.A., Wang J.Y., Jung J.W., Mace D., Pointer S., Barron C., Brady S.M., Schiefelbein J., Benfey P.N. Cell Identity Mediates the Response of Arabidopsis Roots to Abiotic Stress. (2008) Science.

Benfey, P.N., Bennett, M., Scheifelbein, J. (2010). Getting to the root of plant biology : impact of the Arabidopsis genome sequence on root research. The plant journal. **61** : 992-1000.

Handelsman J. (2004). Metagenomics : Application of Genomics to Uncultured Microorganisms. Microbiol. Mol. Biol. Rev., **68** : 669-685.

Hanada S. (2003). Filamentous anoxygenic phototrophs in hot springs. Microbes. Environ, **18** : 51-61.

Chen K. and Pachter L. (2005). Bioinformatics for whole-genome shotgun sequencing of microbial communities. PLoS Comp Biol **1**(2) : 24.

Schloss P.D. and Handelsman J. (2003). Biotechnological prospects from metagenomics. Curr. Opin. Biotechnol. **14** : 303-310.

Riesenfeld C.S., Goodman R.M. and Handelsman J. (2004). Uncultured soil bacteria are a reservoir of new antibiotic resistance genes. Environ Microbiol. **6** : 981-989.

Susannah G.T. and Edward M.R. (2005). Metagenomics : DNA sequencing of environmental samples. Nature Rev. Gen. **6** : 805-814.

Patrick D.S. and Handelsman J. (2005). Metagenomics for studying unculturable microorganisms: cutting the Gordian knot. Genome Biol. **6** : 229.

He J.Z., Zhu Y.G., Zheng Y.M., Zhang L.M., and Shen J.P. (2007). Methodology and application of soil metagenomics. Chinese Academy of Science (Article in Chinese) **18**(1) : 212-218.

Gabor E., K. Liebeton, F. Niehaus, J. Eck and P. Lorenz. (2007). Updating the metagenomics toolbox. Biotechnol, J. **2**(2) : 201-206.

Singh B., S.K. Gautam, V. Verma, M. Kumar and B. Singh. (2008). Metagenomics in animal gastrointestinal ecosystem : Potential biotechnological prospects. Anaerobe. **14** : 138.

Amann R.I., Ludwig W. and Schleifer K.H. (1995). Phylogenetic identification and in situ detection of individual microbial cells without cultivation. Microbiol. Rev, **59** : 143-169.

Xu J. (2006). Microbial ecology in the age of genomics and metagenomics : concepts, tools and recent advances. Mol. Ecol., **15** : 1713-1731.

Babcock D.A., Wawrik B., Paul J.H., McGuinness L., and Kerkhof L.J. (2007). Rapid screening of a large insert BAC library for specific 16S rRNA genes using TRFLP. Microbiol Methods, **71** : 156-161.

Ghazanfar S. and Azim A. (2009). Metagenomics and its Application in Rumen Ecosystem: Potential Biotechnological Prospects. Pakistan Journal of Nutrition. **8**(8) : 1309-1315.

Robe P., Nalin R., Capellano C. and Vogel T.M. (2003). Extraction of DNA from soil of DNA from soil. Eur. J. soil. Boil. **39** : 183-190.

Rolf Daniel. (2004). The soil metagenome-a rich resource for the discovery of novel natural products. Current opinion in biotechnology. **15**(3) :199-204.

Foster R.C. (1998). Microenviroment of soil microorganisms. Biol. Fertil. Soils, **6** : 189-203.

Ranjard L. and Richaume A. (2001). Quantitative and qualitative microscale distribution of bacteria in soil. Res maicroboil **152** : 707-716.

Torsvik V., Ovreas L. and Thingstad T.F. (2002). Prokaryotic diversity-magnitude, dynamics, and controlling factors. Sci. **296** : 1064.

Rosello M.R. and Amann R. (2001). The species concept for prokaryotes. FEMS. Microboil. Rev. **25** : 39-67.

Ovreas L. (2000). Population and community level approaches for analysing microbial diversity in natural environments. Ecol Lett, **3** : 236-251.

Doolittle W. (1999). Phylogenic classification and the universal tree. Science **284** : 2124-2128.

Torsvik V. and Ovreas. L. (2002). Microbial diversity and function in soil : from genes to ecosystems. Curr Opin Microbiol **5** : 240-245.

Curtis T.P. and W.T. Sloan. (2004). Prokaryotic diversity and its limits : microbial community structure in nature and implications for microbial ecology. Curr Opin Microbiol, **7** : 221-226.

●●●

8

Protection of Fruit Growing Soils from the Threats on the Mark

ANIL SHARMA, SARVENDRA KUMAR, NAZEER AHMED, O.C. SHARMA AND SOVAN DEBNATH

There is a need to effectively protect the fruit growing soil and its resources for satisfying elevating fruit demand through natural resource conservation and soil protection strategies, so that the different functions of soil may not be subdued. According to WEH Blum (2005), soils have six different functions which can be distinguished under *Ecological functions* and *Socio-economic functions*.

Ecological functions of soil include :

- Production of biomass, ensuring food, fodder, renewable energy and raw material.
- Filtering, buffering and transformation capacity.
- Biological habitat and gene reserve, with a large variety of organisms.

Where as socio-economic functions include :

- As a basis of technical, industrial and socio-economic structures and their development.
- As a source of raw material *e.g.* clay, sand, gravel, minerals etc.
- As a geogenic and cultural heritage.

These ecological as well as socio-economic functions get tampered in unprotected soils. Unprotected soils need more management than protected soils. But what does soil protection mean or from whom soil needs to be protected. Sometimes soil protection is also considered as one of the soil management strategy but in fact it is merely a tactics to lessen the management efforts efficiently.

Major threats to orchard soils can broadly be divided under five heads:

1. Diminution

(*i*) Fertility depletion
(*ii*) Organic carbon decline
(*iii*) Microbial biomass decline

2. Corporal losses

(*i*) Land Slides
(*ii*) Erosion
(*iii*) Mudslides, Flash floods and Debris flow

3. Contamination

(*i*) Pollution
(*ii*) Aerosol deposition

4. Indefensible land use

(*i*) Unproductive land use
(*ii*) Over grazing

5. Anthropogenic activities

(*i*) Concrete construction and sealing
(*ii*) Over ploughing

1. DIMINUTION

Diminution means decrease or reduction or lessening or shrinking of soil natural resources such as soil nutrients, organic carbon content, soil microbial population etc. This is one of the major problems that the fruit growing soils are introduced with and hence orchard soil needs to be protected from such appraising snag.

(*i*) *Fertility depletion*

One way amputation *i.e.* only removal and no replacement of components which contribute to fertility leads to fertility depletion of soils. Inadequate soil management can be

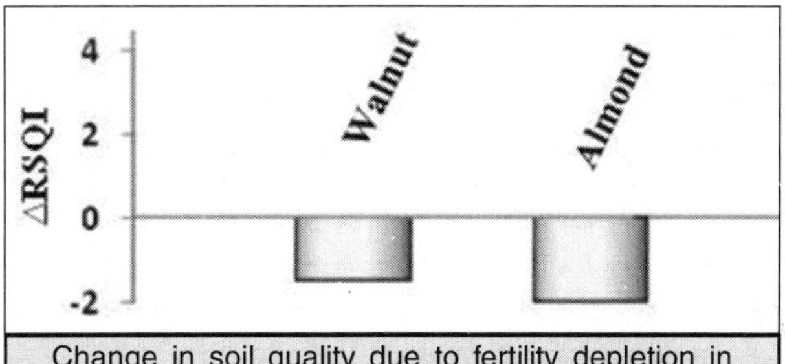

Change in soil quality due to fertility depletion in temperate soils under different fruit crops

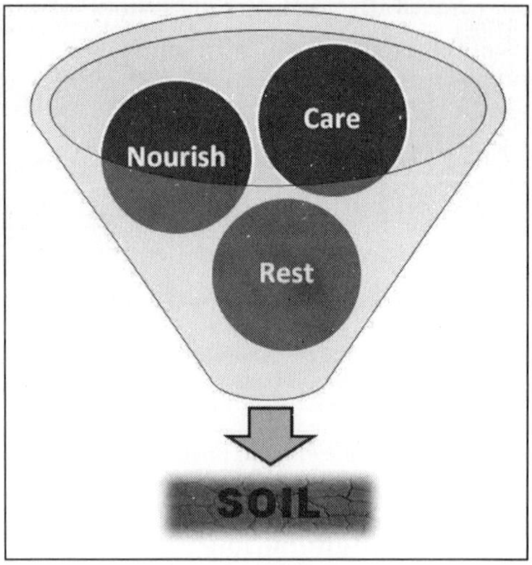

one of the main reasons for such depletion. In times gone by, many past civilizations' collapses can be credited to the soil fertility depletion. In fruit growing soils depletion may occur because of unmanaged or over tillage that shatters soil structure or due to inefficient use of nutrient inputs that escorts to confiscation of the native soil nutrients. Hence this depletion ultimately affects soil quality. Change in relative soil quality index (RSQI) as a result of fertility depletion in fruit growing temperate soils is reflected in figure above.

Managing fertility depletion. Managing, maintaining or escalating soil fertility is one of the most imperative things that need to be kept under continuous consideration to get optimum fruit yields. Sustainable practices and methods to optimize and conserve fertility components should be adopted. These include nutrient application through organic sources such as compost, manure, crop residues; fertilizer trees, intercropping with legumes, fallowing etc. Hence in order to maintain fertility three major mechanisms should always be considered for fruit growing soils and these are nourishment of soil, care of soil and rest to soil. Besides this, management practices to maximize nutrient cycling and nutrient use efficiency, in order to get efficient use of all nutrient sources, should be followed and the basic confront to sustain soil fertility should be overcome.

Basic confront that needs to be met to sustain soil fertility :

- Loss of nutrients should be trimmed down.
- Soil nutrient retention should be elevated.
- Recycling of plant nutrients should be encouraged.
- Adept amount and mode of nutrient application should be adapted.
- Nutrient use efficiency should be escalated.

(ii) Organic carbon decline

Organic carbon decline affects three major aspects of soil fertility *i.e.* chemical, physical and biological. It affects nutrient availability (chemical aspect), soil structure and its related processes (physical aspect) and soil microbial biomass (biological aspect).

Soil carbon amounts are generally related to climatic factors like rainfall and temperature, other factors like vegetation, soil type etc. When the areas get cleared of vegetation all these factors get affected and consequently results in organic carbon decline. Globally, clearing natural vegetation has resulted in large reduction in soil organic carbon levels and further declines may occur due to inefficient management practices (Chan, Y. 2008). Some management practices such as intensive tilling, stubble burning and removal, grass burning can reduce soil

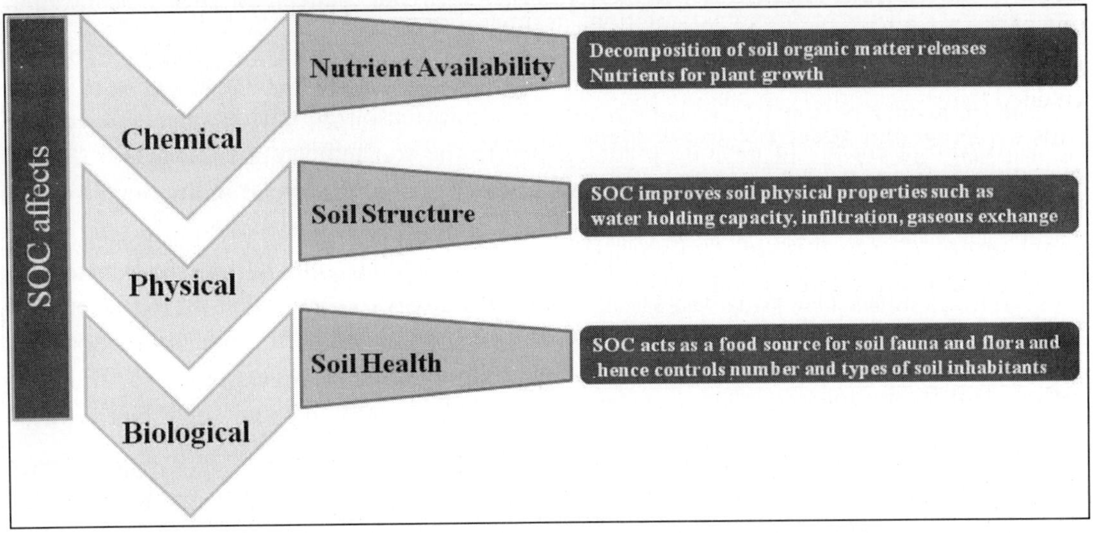

organic carbon by reducing inputs to the soil. Low organic matter addition coupled with intensive tilling can also be one of the reasons for organic carbon decline.

Managing soil organic carbon decline. Soil organic carbon decline can be abridged either by increasing inputs or decreasing losses. There is a wide range of management practices that can increase organic carbon in soils. These practices include improved rotations, improved cultivars, stubble retention, legume cultivation etc. In fact all the practices that increase production in one way or other increases soil organic carbon. Some of the management practices that can increase soil organic carbon content are dealt in the table below.

TABLE 8.1

Management practices that can increase soil organic levels

Management Category	Management practices to increase soil organic carbon
Crop Management	Soil fertility enhancement
	Better rotation
	Erosion control
	Irrigation
Conservation tillage	Stubble retention
	Reduced tillage
	No tillage
Organic amendments	Animal manure
	Green manure
	Recycled organics
Others	Earthworm introduction
	Introduction of legumes

(***Source :*** Chan, Y. 2008)

(*iii*) *Microbial biomass decline*

In the temperate climate of Kashmir valley where temperature falls down upto –10 ± 2°C during winter season, there were several possible reason abrupt decline in microbial biomass. Abrupt change in osmotic potential before the soil temperature exceeds 0°C may be the deleterious consequences of for microbial activity. Loss of psychrophilic organisms as the temperature rises may also contribute to the decline in microbial biomass in cold soils, as may a reduction in substrate availability, and the potentially disruptive effects of freeze thaw cycles. Climate change is expected to cause the loss of an extended period where the soil temperature is sub-zero. In addition, any decline in plant primary production associated with climate change, such as increased summer drought, may be expected to limit the availability of substrates in soils in the following autumn and winter for microbial growth.

Managing microbial biomass decline. Efficient soil management has a major impact on microbial biomass in soils (irrespective of land use pattern and climate), due to its ability to impact the amount of organic carbon entering the soil. The amount of labile carbon present in soil is of particular importance for the biomass. Labile carbon is carbon that is easily broken down by microorganisms, and is largely made up of crop residues and particulate organic matter. This carbon provides a readily available energy source for microbial decomposition, and soils with more labile C tend to have a higher microbial biomass. Management practices that can increase soil carbon and thus soil microbial biomass include :

- The introduction of pasture leys or green manure crops into cropping rotations. These practices add a "pulse" of organic material to the soil that stimulates microbial growth.
- Reducing the length of fallow periods. Fallowing to recharge soil moisture reserves significantly reduces soil biological activity because of the lack of substrate available to soil organisms during this period.
- Retaining crop residues, practices that minimise soil break-up, such as zero tillage, can also help to maintain microbial biomass.
- Maintaining good soil pH-soils that are too acidic can limit microbial activity.

2. Corporal losses

Corporal losses mean physical losses. The main factors responsible for such type of soil losses are slope, rainfall pattern, soil type, topography, crop system and management practices. Corporal soil loss is one of the foremost problems that need to be managed.

(*i*) *Land Slides*

A land slide is a geological phenomenon which includes mass movement of rock, soil and other material going down the slope. Landslides are more likely to occur in areas of land steeper than 45 degrees, though the action of gravity is primary driving force. Main factors responsible for land slides are earthquake; climatic factors mainly rainfall, less vegetation and other anthropogenic activities. Landslides can be categorized based on size (small and large) type of matter (rock, debris and earth) and movement (flows, topples and slides).

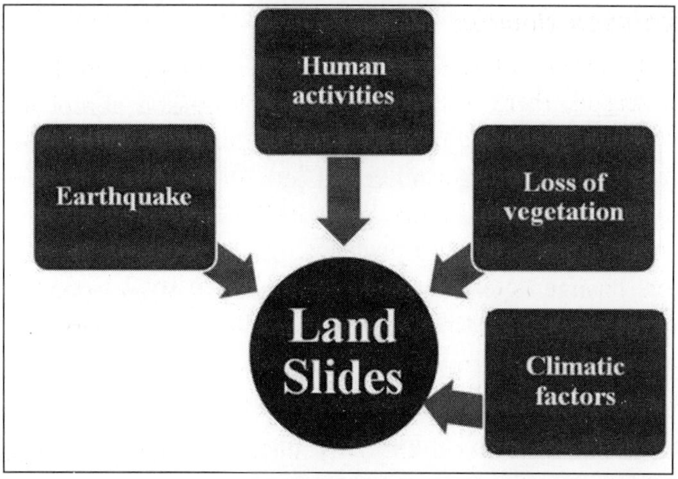

Managing Land Slides. For management it is necessary to find out the reason and then to fix it accordingly. Though landslides are difficult to control, however certain management practices *e.g.* utilizing land based on its capability, planting trees and keeping the soil under vegetation can be adopted to lessen its severity.

(*ii*) *Erosion*

Erosion is the process by which soil and rock are removed from the Earth's surface by exogenetic processes such as wind or water flow, and then transported and deposited in other locations. The main agents of erosion are water, wind and gravity. Erosion is a natural process but is often intensified by human land use practices. While erosion is a natural

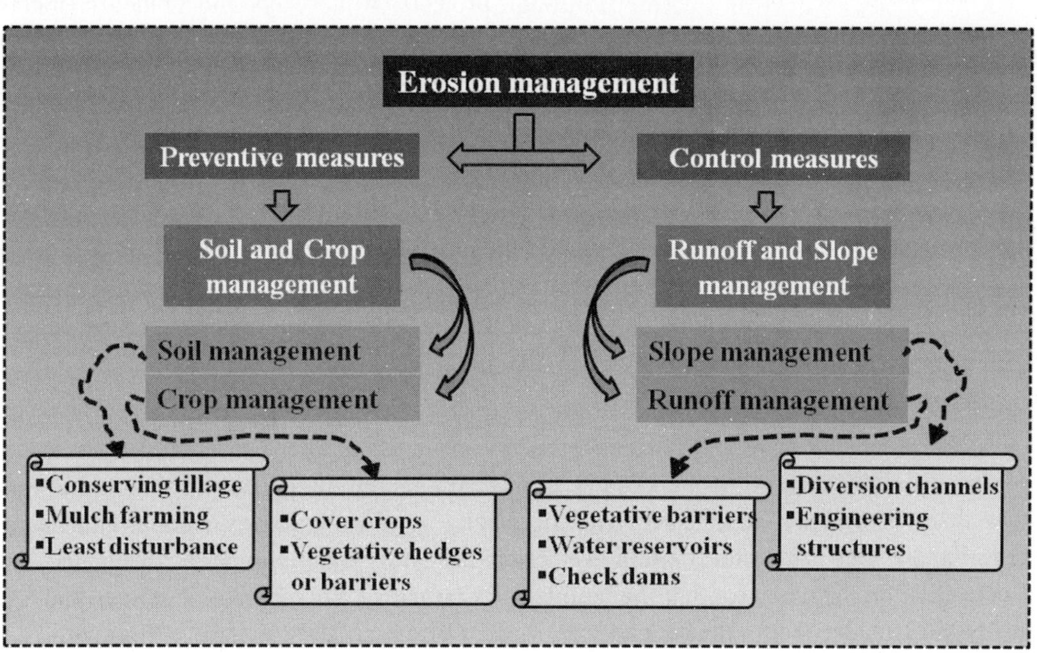

process, human activities have increased by 10-40 times the rate at which erosion is occurring globally. Excessive erosion causes problems such as desertification, decreases in agricultural productivity due to land degradation, sedimentation of waterways, and ecological collapse due to loss of the nutrient rich upper soil layers. Water and wind erosion the two primary causes of land degradation account for 84% of degraded acreage, making excessive erosion one of the most significant global environmental problems.

Managing Erosion. Erosion can be managed by adopting any of the two basic measures : Preventive measures or Control measures. Erosion intensity based management (EIBM) measures/methods should be selected.

Preventive measures include soil and crop management where as control measures include runoff and slope management. Soil can be prevented from erosion through low tillage, mulch incorporation and by minimising the soil disturbance. Crop management to reduce erosion includes cultivation of cover crops and establishment of vegetative hedges or barriers. Runoff and slope management are the two control measures for controlling erosion. Runoff management includes building of vegetative barriers, water reservoirs and check dams. Diversion channels and other engineering structures against the slope are included in slope management measure. In fact effective erosion management lies in (http//:archive.unu.edu) :

- Preventing or minimizing the raindrop impact, through mulching and canopy cover.
- Maintaining favourable soil structure for reducing crusting.
- Managing soil surface to enhance infiltration rate.
- Reducing slope length to minimize runoff build-up.
- Disposing excess runoff safely through protected waterways and graded channels.

(iii) Mudslides, Flash floods and Debris flow

Mudslides are generally the sliding/movement of a large mass of mud formed from loose dirt and water. Some of the mudflows are viscous and therefore slow and some are quick and continuous. These are composed of at least 50% silt and clay-sized materials and up to 30% water. Main causes of mudslides are unusual heavy rains or a sudden thaw. When a mudslide occurs it is given four named areas, the 'main scarp', in bigger mudslides the 'upper and lower shelves' and the 'toe'. The main scarp will be the original area of incidence; the toe is the last affected area(s). The upper and lower shelves are located wherever a large dip (due to mountain or natural drop) in the mudslides path. Most vulnerable mudslide affected fruit growing areas are the orchards at the bottom of slope of hills. Mudslides can destroy a big orchard in no time.

Generally flash floods occur when precipitation falls rapidly on saturated soil or dry soil that has poor absorption ability. The runoff collects in gullies and streams and, as they join to form larger volumes, often forms a fast flowing front of water and other sediments. Fruit growing soils on the slopes and at the banks of the rivers are the most vulnerable to the flash floods.

Flash Flood

Mudslides

Debris flow

A debris flow is a moving mass of loose mud, sand, soil, rock, water and air that travels down a slope under the influence of gravity. To be considered as a debris flow the moving material must be loose and capable of "flow". Debris flows differ from slides because they are made up of "loose" particles that move independently within the flow. A slide is a coherent block of material that "slides" over a failure surface. Mudslides are composed of mud and water. Debris flows have larger particles-at least 50% of a debris flow is made-up of sand-size or larger particles. Orchards at slopes are more susceptible to such type of disaster.

Managing Mudslides, Flash floods and Debris flow : As Mudslides, Flash floods and Debris flow are generally confined to areas down slope and down valley from their points of origin; there should be diversion channels to divert such type of slides and flows. Engineering structures can also be constructed on near the orchards which are prone to such type of disasters. . Such constructions are considered to be a last resort because they are expensive to construct and require commitment to annual maintenance. There should be early warning systems too. In order to prevent debris flows, debris basins may be constructed. Debris basins are designed to protect soil and water resources or to prevent downstream damage.

3. Contamination

Soil contamination is caused due to the accumulation of chemicals or other alterations like waste disposals etc. in the natural soil environment. The ecological system of orchards gets affected due to contamination of soil. The fertility slowly diminishes making land unsuitable for fruit crops. The death of many soil organisms due to contamination can escort to alteration in soil structure too. Pollution because of industrial activities and over fertilization are the main causes of soil contamination.

(*i*) *Pollution*

Soil pollution, in context of orchards, is defined as the accumulation in soils of importunate toxic compounds, chemicals, salts, radioactive materials, or disease causing agents, which have adverse effects on plant growth and consequently the fruit yield. Major causes of soil pollution are industrial waste, agricultural waste, acid rains and other waste disposal. Industrial activity has been the chief donor towards the soil pollution as during the last few decades manufacturing has increased. The contaminated by products are not being disposed off in a way that can be considered safe. As a result, the industrial waste remains in the soil and makes it unbefitting for use. Furthermore, in order to get more and more yield the use of chemical fertilizers and pesticides has increased tremendously. These are full of chemicals that are not produced in nature and cannot be broken down by it. As a result, they seep into the soil after they mix with water and slowly reduce the fertility of the soil. It has also been observed that fruit growers, especially in the hilly areas, are over-fertilizing their orchards which is as bad or worse as under-fertilizing is. The severity of damage caused by over-fertilization is highly variable depending on the type of fruit crop involved. Pests, diseases and even root damage can result from over-fertilization. Dispose of waste is another concern that is considered these days. It has been noticed very frequently that some of the urban sewer system ends in the orchards and causes pollution of soil. Acid rain is caused when pollutants present in the air mixes up with rain and fall back on the soil. This can cause harms to the orchard soils as well as fruit trees.

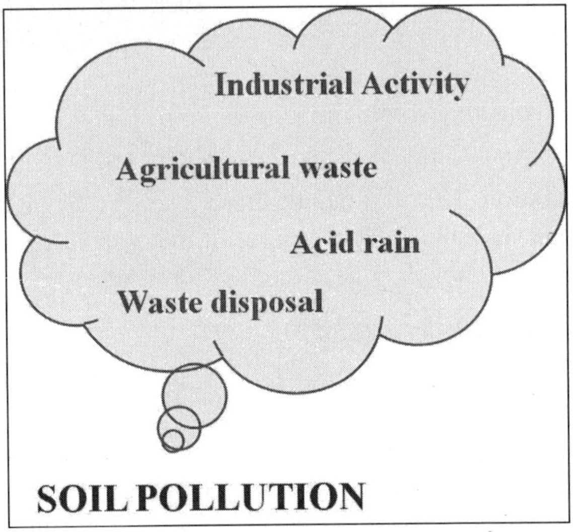

Managing Pollution. Managing soil pollution in fruit growing soils is a much needed demand these days. The pollution due to industrial activity can be managed by installation of proper waste disposal units. Although there is law to check the soil contamination through these wastes but these need to be implemented and monitored properly. Use of excessive chemicals in agriculture as well as horticulture should be avoided. Use of slow release fertilizers as well as organic manures should be promoted. There should be waste treatment plants just on the periphery of urban areas so that the contaminants may not enter the soil.

(*ii*) *Aerosol deposition*

There are various types of suspended particulate matters in air such as soil dust, sea salt, sulfate from sulfuroxides, nitrate from nitrogen oxides etc. These are called aerosols An aerosol is a colloid of fine solid particles or liquid droplets, in air. Examples of aerosols include haze, dust, particulate air pollutants and smoke. The liquid or solid particles have diameter mostly smaller than 1 mm or so (Toshihiko Takemura, YN*). These aerosols, even if in air, causes significant damage to fruit growing soils and crops, whether directly through deposition or indirectly by effecting the environment. Aerosols scatter and absorb the thermal radiation that is emitted from the Earth's surface and atmosphere and the solar radiation, leading to a change in radiation balance. The effect is much dependent on the physical and chemical properties of aerosols, such as particle size, hygroscopicity, and complex refractive index (Toshihiko Takemura, YN*). Aerosol deposition affects soil flora and fauna. The microbial biomass gets disturbed and so is the nutrient release and its availability. Aerosols are responsible for changing the micro-climate too.

Managing aerosol deposition. Aerosol deposition can be managed by reducing the pollution. This will reduce the aerosols in the air and subsequently its effect on soil flora and fauna as well as micro climate of the orchards. Besides the aerosol emitting industries should not be established near the fruit orchards.

4. Indefensible land use

There is a need to explore defensible land use options, especially for fruit growing soils to reduce the risks of total crop failure due to increase in extreme climatic events. Population increase leading to uncontrolled land use patterns which is ultimately affecting soil fertility. There is a need to encourage fruit planting as per the land suitability for that fruit crop and not on the basis of demand of a particular fruit crop, as it leads to unproductive land use.

(*i*) *Unproductive land use*

High fruit demand due to their nutritional values has resulted in annual addition to total area under these, but this expansion has not been in proportion to production. This disproportion is as a result of unproductive land use system. Unsystematic as well as unproductive land use planning and management practices (Sharma *et al.,* 2010) can be considered as culprits for such a misappropriation. The production of fruits is greatly influenced by a number of factors which include climatic factors like rainfall (Dorfman *et al.,* 1988) and temperature; soil factors like fertility (Dominguez *et al.*, 2010) and available nutrients (Nyomara and Brown, 1999); geographical factors; physiographic and varietal factors (Lamp *et al.*, 2001) etc. So the land use should be planned on the basis of production efficiency of a particular fruit crop under certain climatic, soil, physiographic and geographic conditions. For example if some area is suitable for a particular fruit crop should not be used for a fruit crop for which the area is less suitable. Here suitability has its direct relation to production, productivity and quality. There are a number of methods suggested for proper land use especially in case of fruit crops. One of these is production efficiency based land use planning.

Managing unproductive land use. For managing unproductive land use there is a need to develop certain indices and models which can perform the preliminary production efficiency analysis of different locations on which the orchard plantation is being planned. One of such model/software application (Predictor and Planner for Almond) for almond has already been developed by Central Institute of Temperate Horticulture, Srinagar. Unproductive land use can also be managed by good planning and management. For example Land considered "unproductive" might not be profitable for growing one type of fruit crop but there is a possibility that this land acts as productive for another fruit crop.

(*i*) *Over grazing*

Grazing within the orchards can easily be seen at many places especially in northern India. Grazing is generally not considered as a big problem in relation to orchard soil but this grazing when turns to overgrazing becomes one of the key factors for land degradation. It is important that vegetation covers the ground to protect orchard soils from exposure. Overgrazing takes away this protective vegetation, whereas livestock hooves flatten exposed soils. These soils are then susceptible to erosion, which remove nutritionally rich upper layers of the soil. The process of land degradation further speed up under water stress conditions. Overgrazing effects the fruit growing soils in two ways: Reduction and Induction. It reduces microbial biomass, water quality and organic carbon; at the same time it induces erosion, nutrient loss and root exposure. When livestock clip grass by grazing, the plant is injured. Because the shoots have been clipped and eaten they are no longer able to perform their function of collecting energy from the sun. The plant must tap into the energy reserves in the

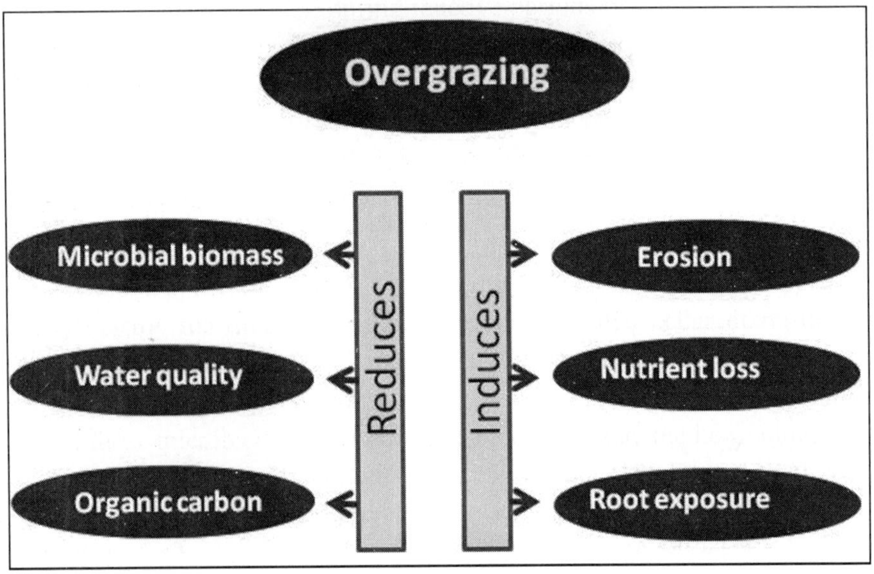

root in order to stay alive and begin regrowth. This means the roots will now die back as the energy stores are used up to regrow shoots (http://www.ranching-with-sheep.com/overgrazing.html).

Managing overgrazing. Avoiding overgrazing and maintaining vegetation cover in orchard soils is the only way to avoid losses caused by overgrazing.

5. Anthropogenic activities

The way we people use land in and around fruit growing areas affect the soil by altering various soil processes and resources. Human activities like constructions, chemical applications, over ploughing etc. to get maximum return per unit area, has resulted in swift decline in soil quality of orchards. These anthropogenic activities need to be given consideration in order to protect fruit growing soils.

(i) Concrete construction and sealing

The current rate of soil degradation has increased due to concrete constructions because of urban expansion. Many concrete infrastructures can be seen alongside the orchards. These infrastructures, if large, prevent sunlight to reach the orchard soils and thus affect various soil biological processes. Besides this these infrastructures affect soil aeration and consequently nutrient availability too. Sometimes improper drainage also results due to such structures. Here in such orchards ground water quality gets affected due to leaching of various harmful chemicals coming out from these heavy infrastructures *i.e.* factories etc.

Managing soil degradation due to concrete constructions. Well planned concrete constructions or no construction alongside the fruit growing soils can protect soils from degradation. Strict laws should be made so that orchard soil degradation can be avoided.

(ii) Over ploughing

The primary purpose of ploughing in orchards is to bury down the weeds. But over ploughing orchard soil can affect the soil badly. Over ploughing makes soil susceptible to erosion, high soil moisture loss, fuel loss, labour costs, increased dependency on chemical fertilizers and slow soil warming especially in poor drained temperate soils. Besides this it causes injury to roots, nutrient loss and disturbs microbial population.

Managing over ploughing. Over ploughing or heavy ploughing should be avoided in orchard soils. Only light ploughing, that too when needed, should be done.

REFERENCES

Blum WEH. (2005). Functions of Soil for society and the environment. Reviews in Environmental Science and Biotechnolgy. **4** : 75-79

Chan Y. (2008). Increasing soil organic carbon of agricultural land. Primefact 732. NSW Department of Primary Indistries,Wagga Wagga.

Dominguez, S.G., Faz, A., Zornoza., (2010). Characterization of almond orchards to assess soil fertility and organic matter dynamics to improve soil conditions by organic amendments. 19th World Congress of Soil Science, Soil solutions for a changing world 1-6 August, 2010, Brisbane, Australia pp. 39-42.

Dorfman, J., Dorfman, M., Heien, D., 1988. Causes of almond yield variation. California Agriculture, 27-28.

Lamp, B.M., Connell, J.H., Duncan, R.A., Viveros, M., Polito, V.S., (2001). Almond flower development: floral initiation and organogenesis. Journal of American Society of Horticulture Science **126**(6) : 689-696.

Nyomora, A.M.S., Brown, P.H. (1999). Rate and time of boron application increase almond productivity and tissue boron concentration. Hort Science **34**(2) : 242-245.

Sharma, A., Jalali, V.K., Arora, S., (2010). Non-exchangeable potassium release and its removal in foot-hill soils of North-west Himalayas. Catena **82**(2) : 112-117.

http://www.ranching-with-sheep.com/overgrazing.html

http//:archive.unu.edu (VII Runoff management and erosion control)

●●●

9

Maintaining Soil Microbial Balance in Fruit Growing Soils—A Challenge

SARVENDRA KUMAR, NAZEER AHMED, ANIL SHARMA AND O.C. SHARMA

Soil is an essential component of the ecosystem and it is the result of the mineral, chemical, physical and biological components present in the soil. The understanding and maintaining biological processes that take place in the soil is crucial for efficient soil use and to preserve soil health. The biological component of the soil is mainly represented by microorganisms and enzymes associated with them, which are involved in numerous processes such as different nutrient cycling, soil organic matter decomposition, soil formation, therefore they are important for the maintenance of the soil health and for plant productivity, therefore has, economic as well as ecological implications. Microorganisms can be affected by abiotic factors such as temperature, moisture, seasons and soil nutrients availability, or by biotic factors, namely interactions with other microorganisms, plant etc.

Therefore maintenance of microbial communities in soil especially in horticultural crop under is challenge as well as new concept for soil microbiologist. The quantifications of microbial communities from soil, however, is a subject of continuous debate, which seems to be caused by "its definition, which includes non growth components other than maintenance; the existence of partly overlapping concepts; the evolution of variables as constants; and the neglect of cell death in microbial dynamics". Over the years, microbial maintenance in soil has been quantified in various, partly contradictory, ways. Maintenance has been defined as "*the energy consumed for functions other than the production of new cell material*". This definition includes all non growth components and does not provide insight in its underlying processes, which complicates its general application. Physiological maintenance comprises energy costs of osmoregulation, cell motility, defense mechanisms, and proofreading and internal turnover of macromolecular compounds. But, it should not include shifts in metabolic pathways, storage of polymers, or extracellular losses (Chesbro, 1988). Some of the physiological maintenance components, such as shifts in metabolic pathways seem to be better in place when corrected for in the cell growth yield (Y_G), which was defined as the amount of cell growth per amount of substrate consumed. The most popular equation used to describe microbial growth, the *Monod equation* (Monod, 1949), was introduced as a purely empirical relationship resembling an adsorption isotherm. However, many microbiologists viewed the Monod equation as something that has an inherent meaning. From there, it is a small step to consider the variables involved as constants for the species considered. When measurements of maintenance showed deviations from a constant maintenance, the concept itself was not questioned, but an additional growth rate dependent maintenance parameter was introduced (Pirt, 1982). Similar evolution occurred with growth yield Y_G and the energy yield (YATP), which was introduced based on a correlation between biomass production and ATP availability. These were subsequently proven to be constants, whereas Stouthamer *et al.* (1990) concluded that Y_G is not a biological constant, based on measured shifts in metabolic pathways.

ESTIMATORS FOR MAINTENANCE

Specific Maintenance Rate (endogenous metabolism): *"negative relative growth rate, in which the sum of the specific maintenance rate times microbial biomass equals the loss of cell material through maintenance"*

$$\frac{dx}{dt} = \mu x - ax \qquad \ldots(i)$$

$$-\frac{dS}{dt} Y_{app} = \frac{dx}{dt} \qquad \ldots(ii)$$

where x denotes microbial biomass (in g C_x m^{-3}), m is the true specific growth rate (in s^{-1}), a is the specific maintenance rate (SMR) (in s^{-1}), S is the energy supplying substrate (in g C_s m^{-3}), and Y_{app} is the apparent yield coefficient (in g C_x g^{-1} C_s). Y_{app} describes the gross partitioning of elements over cell biomass and extracellular products.

As shown in equation(1), the definition of SMR (a) is completely analogous to the definition of true specific growth rate (m). Maintenance is an imaginative decay rate to account for a diversion of substrate flux from growth, but does not necessarily lead to additional losses of substrate. In fact, a includes osmoregulation, extracellular losses and turnover terms, and neglects the occurrence of the other non-growth components like energy spilling reactions, cell motility, changes in stored polymeric carbon, shifts in metabolic pathways etc. Given its mathematical definition, a also includes relative death rates when a is determined experimentally from biomass and compound balances, *e.g.*, in chemostat experiments. Estimates derived for either SMR or relative death rates thus overlap. It also implies that SMR includes non-growth components other than physiological maintenance, whereas some physiological maintenance processes are not incorporated. More complicated models are needed to estimate real physiological maintenance rates.

The general definition given above does not imply that *a* is a constant. However, the empirical relation developed Marr *et al.* (1963) to estimate a treated *a* as a constant :

$$\frac{1}{x} = \frac{a}{x_{\max}}\frac{1}{\mu} + \frac{1}{x_{\max}} \qquad \ldots(iii)$$

in which X_{max} was incorrectly assumed to be a constant maximum microbial biomass (in g C_x m^{-3}). Still, most of the models that apply *a* and treating it as a constant, although a constant *a* does not explain the persistence of microbial communities under low nutrient conditions. Moreover, experimentally derived values for *a*, assuming its constancy, for different microbial species shows a significant correlation with its x_{max} value ($P < 0.001$; correlation coefficient 0.73; Fig. 9.1). This indicates a higher SMR (*a*) for faster-growing micro-organisms. There is also evidence that SMR (*a*) decreases at decreasing μ within microbial species and even more strongly at starvation also pointed out by Russell *et al.* (1995).

Specific maintenance rate (*a*) depends upon growth conditions; but direct application of its values should be done with care and only under conditions similar to those at which it was determined. Alternatively, SMR may be made explicitly dependent upon growth conditions and growth history. In a theoretical study, Powell (1967) proposed that the instantaneous specific substrate consumption rate is related to the organism's activity called its physiological state. This concept of physiological state *r* was applied and incorporated in a Synthetic Chemostat Model (Pnaikov, 1995). By definition, *r* varies between 0 and 1, where $r = 1$ represents maximum activity and $r = 0$ represents complete dormancy. Consequently :

$$a = ra_{\max} \qquad \ldots(iv)$$

where a_{max} is the maximum specific maintenance rate (occurring if $S \geq \infty$). The complication is to derive *r* and its dynamics. Panikov (1995) proposed for the steady state *r* value, r_r :

$$r_r = \frac{S_r}{K_r + S_r} \qquad \ldots(v)$$

where K_r is a saturation constant (in g C_s m^{-3}), analogous to Michaelis-Menten constant and the Monod saturation constant, and S_r is the equilibrium substrate concentration (in g C_s m^{-3}). Next to the theoretical foundation of this equation, the parameter estimation of K_r is especially problematic.

Maintenance Coefficient. Maintenance coefficient (*m*) defined by Pirt, (1982) as "*the minimum substrate consumption to maintain the cells (m in g* C_s g^{-1} C_x s^{-1}*)*"

$$q \equiv -\frac{1}{x}\frac{dS}{dt} = m + \frac{\mu}{Y_G} \qquad \ldots(vi)$$

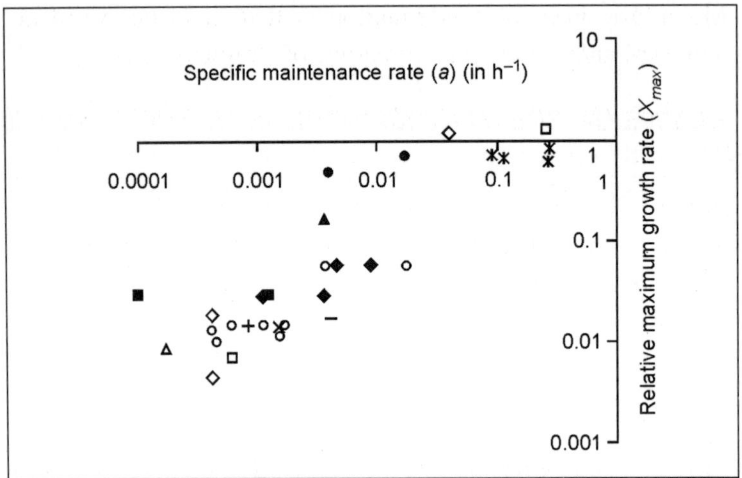

Fig. 9.1. Correlation between measured specific maintenance rates *a* and maximum relative growth rates x_{max} for various microbial species (Peter *et al.*, 2005)

Where, q is the specific substrate consumption rate (in g C_s g^{-1} C_X s^{-1}), used for growth and for maintenance. Maintenance thus denotes extra substrate consumption not used for growth purposes. In this approach, no biomass losses are possible, which is conceptually unattractive and which in fact assume m to be the net relative growth rate. This leads to substrate consumption while substrate concentrations are zero. Defined in this way, m neglects the extracellular losses, osmoregulation, and turnover of macromolecular compounds which is used for maintenance of microbial cell in soil.

In its original definition, the maintenance coefficient (m) was considered constant for a species and has been used as such in numerous models describing microbial dynamics. However, in analogy to the understanding that a depended on growth conditions, it was reported by several researcher that the maintenance coefficient (m) depended on m [4, 7, 11, 32]. Therefore, on the assumption that the maintenance coefficient includes a portion that decreases with increasing m, and relationship between them were further postulated by Pirt, (1982).

$$q = m_c + \frac{\mu}{Y_G} + m_v \left(1 - \frac{\mu}{\mu_{\max}}\right) \qquad \ldots(vii)$$

where m_c denotes the constant maintenance coefficient (in g C_s g^{-1} C_X s^{-1}) and m_v $(1\text{-}m/m_{\max})$ denotes the growth rate dependent maintenance coefficient (in g C_s g^{-1} C_X s^{-1}). It is important to distinguish Y_{app} from Y_G, which is the yield coefficient corrected for maintenance. In other words, Y_G considers only consumption of substrate for growth purposes (in g C_X g^{-1} C_s). Given the correction for maintenance, Y_G is supposed to be higher and less variable with S than Y_{app}. Equation (*vii*) incorrectly assumes that maintenance is only a function of time and biomass. Indeed, these mathematical constants are not biological constants, but both may vary with *m*. Major drawback of this equation is that, it is not valid at very low *m* and attributed this not explain/account the formation of dormant cells.

RELATION BETWEEN SPECIFIC MAINTENANCE RATE AND MAINTENANCE COEFFICIENT

Several authors have tried to link the two maintenance estimators. Inconsequent application of the two above mentioned yield parameters, however, have led to an erroneous derivation of a relationship between *m* and *a*. From equation (*vii*), in combination with the definitions of Y_G and Y_{app}, it can be derived that :

$$\frac{1}{Y_{\text{app}}} = \frac{\mu}{Y_G} + \frac{m}{\mu} \qquad \ldots(viii)$$

Similarly, from equations (*i*) and (*ii*), in combination with the definitions of Y_G and Y_{app}, it follows that :

$$\frac{1}{Y_{\text{app}}} = \frac{\mu}{\mu - a}\,\frac{1}{Y_G} \qquad \ldots(ix)$$

Equation (*ix*) predicts that Y_{app} decreases with decreasing growth rates if both Y_G and *a* are assumed constant, which indeed had been proved experimentally. Substitution of $1/Y_G$ in either Equations (*viii*) and (*ix*) shows that :

$$a = mY_{\text{app}} \qquad \ldots(x)$$

which is only equal to m = a/Y_G if $Y_{app} = Y_G$. It follows from Equations (*viii*) and (*ix*) that this only happens to be true if maintenance is absent/or, in other words, if *m* [Equation (*viii*)] is zero or, following the alternative approach, if *a* [Equation (*ix*)] is zero.

Equation (*x*) implies that there is no constant relation between the two parameters describing maintenance (Fig. 9.2), because Y_{app} depends on *m* if maintenance occurs. Without sophisticated experimental tools, it is impossible to tell which parameter is non-constant, because the two approaches on maintenance address partly the same and partly different non-growth components and even cell death, whereas shifts in metabolic pathways are neglected by both approaches. For each component, the contribution to overall non-growth losses may shift both in absolute and in relative terms with growth conditions. Therefore, the two ways of describing "*maintenance*" should be seen as two independent measures approaching a complex phenomenon from different directions. Combining empirical parameter estimates from both approaches in one model should thus be done with great reservation and care. In reality, it is highly probable that given the different components combined in "*maintenance*" partial losses occur in both biomass and in consumption.

Only a few studies tried to separate the maintenance estimators. Servais *et al*. Year quantified the relative release rate of label that had been incorporated into DNA as a relative death rate or specific maintenance rate assuming that DNA is not subject to maintenance.

Servais *et al.* (1985) also showed that the temperature dependence of this specific maintenance rate was slightly less than generally found for relative growth rates.

Maintenance as a Minimum Substrate Concentration. *Minimum substrate concentration (S_{min}) above which growth occurs, instead of explicitly accounting for maintenance and/or death* introduced by Jetten *et al.* (1990). Sometimes, S_{min} was linked to the occurrence of reverse enzymatic reactions. Below S_{min}, it is usually assumed that growth is zero although it would be more correct to consider growth negative at concentrations lower than S_{min}. In fact, the introduction of S_{min} shifts the μ–[*S*] curve horizontally, where introducing *a* shifts the μ–[*S*] curve vertically (Fig. 3). Both parameters can be easily converted into each other. By definition, $S = S_{min}$ if $dx/dt = 0$. Assuming Monod kinetics, then

$$a = \frac{\mu_{max} S_{min}}{K_s + S_{min}} \qquad S_{min} = \frac{aK_s}{\mu_{max} - a} \qquad \ldots(ix)$$

where, K_s is the Monod saturation constant (in g C_s m^{-3}) and m_{max} is the maximum relative growth rate (in s^{-1}). In equation (*ix*), *a* simply equals *a* at $S = S_{min}$ and is not necessarily a constant.

In soil, microbial communities maintaining several critical function (Table 9.1), one essential microbial function in soils is the processing and recovery of key nutrients from detritus inputs and accumulated soil organic matter. Besides that, are involved in several complex biogeochemical cycles of nature, which are largely responsible for maintenance of soil fertility and plant productivity. Soil micro-organisms influence plant productivity through mainly two maechnism : direct effects on plants via root-associated organisms that form mutualistic or pathogenic relationships with plants, and indirect effects via the action of free-living microbes that alter rates of nutrient supply and the partitioning of resources.

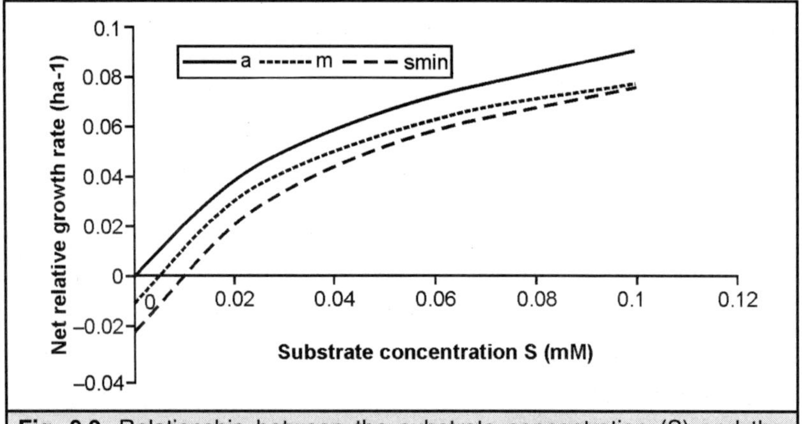

Fig. 9.2. Relationship between the substrate concentration (S) and the net relative growth rate (m) while accounting for maintenance in different ways, *i.e.*, by a specific maintenance rate (*a*), maintenance coefficient (*m*), or a minimum substratem-concentration (S_{min}).

POSITIVE DIRECT EFFECTS

Microbial diversity can also promote plant diversity and productivity when microbes associate with different plant species or when different microbes provide different resources. A wide range of soil microbes form intimate symbiotic associations (*e.g.* plants and nitrogen (N)-fixing bacteria etc.) with plants and can stimulate plant productivity by supplying limiting nutrients to the plants. Nitrogen-fixing bacteria are important regulators of plant productivity because plants cannot fix atmospheric N and nitrogen together with phosphorus (P) and potassium, the main element that limits plant productivity. The contribution of N-fixing bacteria to plant productivity is thought to be biggest challenge in temperate climate, in this situation, N-fixing bacterial symbionts can contribute up to 20% of all plant N that is annually acquired by the vegetation especially fruit crop, through proper management. Several other plants form associations with N-fixing bacteria, including c. 400 shrubs that associate with *actinomycetes*, approximately 150 cycad and 65 Gunnera species that associate with cyanobacteria (Rai *et al.*, 2000), and an unknown number of plant species that harbour endophytes and can fix N.

Another important group of plant symbionts that enhance plant productivity by supplying limiting nutrients are mycorrhizal fungi. Mycorrhizal fungi are widespread and form symbiotic associations with the roots of 80% of all terrestrial plant species (Smith and Read, 1997). Mycorrhizal fungi can provide resistance to disease and drought, and supply a range of limiting nutrients including N, P, copper, iron and zinc to the plant in exchange for carbon. Mycorrhizal fungi often enhance resource complementarity by providing nutrients that are otherwise inaccessible to plant roots. The most abundant and important groups of mycorrhizal fungi are the arbuscular mycorrhizal (AM) fungi, the ecto-mycorrhizal (EM) fungi and the ericoid mycorrhizal (ERM) fungi. Enhanced phosphorus uptake is one of the mechanism by which AM fungi can enhance plant productivity and experiments with single plants (horticultural) communities have shown that AM fungi contribute up to 90% of plant P uptake (Van der Heijden *et al.*, 2006b). Enhanced P uptake is especially important for horticultural plant where phosphorus is low in soil and less microbial activity due to climatic and other factors. Some recent studies have also shown that AM fungi can contribute to enhanced N-acquisition under some conditions (Hodge *et al.*, 2001); however, the significance of AM fungi for N uptake is unresolved. EM fungi associate with c. 6000 tree species and are abundant in temperate and boreal forests and in some tropical forests (Alexander and Lee, 2005). The presence of ecto-mycorrhizal fungi is of pivotal importance for plant productivity in most temperate areas. Nutrient availability in these ecosystems is usually low and most nutrients are present in organic form in litter and humus. Therefore, ecto-mycorrhizal fungi can acquire N from litter through extensive hyphal networks that forage for nutrients and by excreting a wide range of extracellular enzymes that can degrade organic matter.

Recent work has shown that microbial diversity in soil ecosystems is reduced because of land-use intensification and increased nutrient availability (*e.g.* Helgason *et al.*, 1998; Fig. 9.3), nitrogen deposition and chemical contamination (Gans *et al.*, 2005). The impact

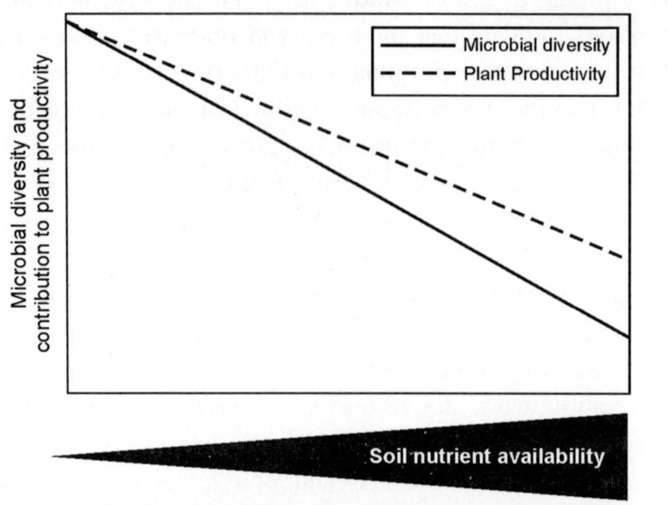

Fig. 9.3. Hypothetical relationship between nutrient availability and the microbial contribution to plant productivity. Microbes are hypothesized to be most important for the productivity of nutrient poor ecosystems. It is also hypothesized that microbial diversity is negatively correlated with nutrient availability.

of this reduced diversity on plant diversity and productivity is unclear. There is one hypothesis that (Heijden *et al.,* 2008) significance of microbes is highest at low nutrient availability (Fig. 9.2) and nutrient poor ecosystems are more vulnerable to reductions in microbial diversity and losses of specific microbial species or functional groups. These expectations are based on several observations. First, plant productivity in nutrient poor ecosystems is often enhanced by a range of microbial symbionts that acquire various limiting nutrients as discussed earlier. Several of these symbionts have a restricted host range (*e.g.* several rhizobia and some mycorrhizal fungi) and a reduction in microbial diversity could reduce growth of plant species that depend on specific microbial symbionts. For instance, Jonsson *et al.* (2001) observed that effects of fungal diversity were strongest at low nutrient availability, perhaps because different mycorrhizal fungi obtained limiting nutrients from different sources in the soil. In contrast, plants are often less dependent on mycorrhizal fungi and N-fixing bacteria when nutrient availability is high (Sprent, 2001).

Second, nutrient poor ecosystems are especially vulnerable to nutrient loss (*e.g.* because of microbial denitrification or leaching) because plant productivity is limited by nutrients in these ecosystems. Thus, small nutrient losses will immediately lead to a reduction of plant productivity in nutrient poor ecosystems. Third, chemical diversity in nutrient poor ecosystems is often high because such ecosystems are usually species rich, containing many different plant species that produce a wide range of secondary metabolites and recalcitrant defence compounds, including lignin and tannins. It is likely that a wide range of physiological diverse microbes contributes to the breakdown of plant litter in nutrient poor ecosystems. Hence, these observations indicate that microbes and microbial diversity are likely to have the biggest impact on ecosystem performance in nutrient poor ecosystems. Following this, it is very clear that the relationship between microbial diversity and ecosystem functioning is different for nutrient poor and nutrient rich ecosystems. Microbial communities from nutrient rich ecosystems are functionally more redundant compared with microbial communities from nutrient poor ecosystems where microbes need specific adaptations to obtain resources [*e.g.* during decomposition or when forming (host specific) symbiotic associations with plants]. Effects of increasing microbial diversity on plant productivity probably also depend on the number of functional groups that are present. Adding increasing numbers of species from a particular functional group (within functional group diversity) will probably add some degree of functional diversity and have slight effects on ecosystem functioning. However, adding different functional groups (between functional group diversity) or specific keystone species will probably have a bigger impact on plant productivity and particular ecosystem processes. The level of functional diversity may also be related to the phylogenetic relatedness among soil microbes. Interestingly, despite the high level of bacterial species diversity found in soil, they have a lower phylogenetic diversity than other environments (Lozupone and Knight, 2007), perhaps suggesting that soils contain a large number of microbial species with similar functions.

Negative direct effects. Microbes can also reduce plant productivity and this can have dramatic/lethal consequences for ecosystem as well soil ecological processes, especially when dominant or keystone species, such as horticultural/forest trees are affected. For instance, soil pathogens including *Phytophthora* spp., *Fusarium* spp. and *Pythium* spp. have been shown to attack a range of horticultural crop including apple, walnut, almond and cherry. It is difficult to estimate the impact of microbial pathogens on plant productivity in natural communities because a reduction of biomass of one species is often compensated by additional growth of other species, especially in species rich plant communities.

Positive indirect effects. Microbes can also indirectly influence plant productivity via the action of free-living microbes that alter rates of nutrient supply and resource partitioning. These effects can either stimulate plant productivity, through the actions of microbes that enhance the availability of nutrients for plant uptake, or reduce plant productivity through competition for nutrients with plant root and/or by promoting nutrient loss via leaching of mobile nutrient forms. Perhaps the most important route by which free-living microbes influence plant nutrient availability, and hence plant productivity, is via processes of nutrient mineralization, whereby soil microbes break down soluble and insoluble organic matter and convert it into inorganic, plant available forms. Most soil N (some 96-98%) is contained in dead organic matter as complex insoluble polymers such as proteins, nucleic acids and chitin, and these polymers are broken down into dissolved organic N (DON) by extracellular enzymes that are produced by soil microbes. This DON, which can constitute a significant portion of the total soluble N pool, is either absorbed by free-living soil microbes, or it is mineralized by the microbial biomass (under conditions when microbial growth is C limited), thereby liberating inorganic-N into the soil environment. Alternatively, plants might take up DON directly from soil, in the form of amino acids, thereby by-passing the microbial mineralization step. This was shown to be the case in many ecosystems, but especially in those that are strongly N limited, as in most of Indian soil. Free-living N-fixing bacteria, which are ubiquitous in terrestrial ecosystems, can also contribute significantly to the N budget of some systems; they fix relatively small, but significant amounts of N (< 3 kg N ha^{-1}year) (Cleveland *et al.*, 1999). Microbial uptake of nutrients also has important implications for ecosystem nutrient storage and temporal partitioning of nutrients between plants and soil microbial pools. Immobilization of N by microbes, for example, has been shown to act as a short-term sink for N in several terrestrial ecosystems, thereby potentially limiting the export of N to adjacent ecosystems and groundwater (Brooks *et al.*, 1998). Seasonal patterns of N immobilization by microbes are also important for plant N-acquisition, especially in strongly N limited ecosystems where microbial communities immobilize N maximally in autumn, after plant senescence, and retain it throughout the winter until spring, when it is released for plant uptake (Bardgett *et al.*, 2005). Microbial uptake of N is also important for longer-term ecosystem N retention, via the transfer of the nutrients from within microbial tissues to more stable organic matter pools after cell death. Other routes by which microbes can alter plant nutrient availability include weathering of soils via the exudation of organic acids and the solubilisation of various forms of precipitated P; on the basis of laboratory assays, it was shown that P-solubilizing bacteria may constitute up to 40% of the cultivable population of soil bacteria (Kucey 1983). The significance of these bacteria in natural ecosystems is still unclear and is an area that warrants further attention. Another route by which microbes affect plant productivity is disease suppression, for example through the production of antifungal metabolites by *Pseudomonas* sp. These bacteria protect several major horticultural crops against diseases, a phenomenon that is likely to be also important in natural ecosystems.

Negative indirect effects

As discussed above, microbes compete with plants for nutrients in soil solution, with possible negative effects on plant nutrient acquisition and growth. This is most likely the case in strongly nutrient limited soil, where soil microbes have been shown to compete effectively with plants for N leading, in some cases, to similar quantities of N being found in the microbial and plant pool. Moreover, it is demonstrated that if microbial activity is stimulated through the addition of labile carbon (C) to the soil, plant N uptake and plant productivity decline (*e.g.* Schmidt *et al.,* 1997). Another route, by which microbes reduce ecosystem N availability, is by transforming N to more mobile forms, such as nitrate which is produced by the bacterial process of nitrification. Nitrate is much more mobile in soil than ammonium and can be lost easily via leaching into ground water and surface run-off. Hence, nitrifying bacteria can indirectly reduce plant productivity because they reduce the availability of a nutrient that is often limiting plant productivity. Substantial amounts of nitrate can be found in groundwater or subsurface water and this can represent a substantial loss of N, especially in temperate humid ecosystems and undulating/sloppy lands. Nitrifiers, other bacteria and some fungi can transform nitrate into N gases (denitrification) under anaerobic conditions. The impact of denitrifiers on plant productivity and plant community composition has rarely been tested. It should be large because up to 50% of available soil N can be lost by denitrification from tropical condition. So, it is important to understand the impact of denitrifying communities on denitrification rates because world-wide, an estimated 105-185 tons of N are lost yearly through denitrification by bacteria (Tiedje 1988). This equals c. 7% of the world-wide terrestrial productivity, assuming that N is limiting plant productivity (Schlesinger 1997). Estimates of microbial effects on plant productivity are often difficult because populations and communities of soil microbes are affected by a wide range of other soil biota, especially their consumers, such as protozoa, Collembola etc. For instance, bacterial

TABLE 9.1

Influence of soil microbes on various ecosystem processes

Ecosystem process	Microbes involved	**Estimated microbial contribution to ecosystem process**
Carbon cycle		
Plant productivity	Nitrogen-fixing bacteria, mycorrhizal fungi	0-50%*
Decomposition	microbial pathogens	–50-0% #
	Bacteria, fungi	Up to 100% !!
Nitrogen cycle		
Plant nitrogen acquisition	Rhizobia, *actinomycetes,* free-living	0-20%$ (sometimes higher)
Nitrogen fixation	bacteria	
Soil uptake Nitrogen loss	Mycorrhizal fungi	0-80%
Denitrification		¶
Enhanced leaching because of nitrification	Denitrifiying bacteria and some fungi	Up to 60%**
		??
Phosphorus cycle		
Plant phosphorus acquisition	Nitrifying bacteria, archaea	0-90% †
Phosphorus loss because of leaching after mineralization	Mycorrhizal fungi, P-solubilizing bacteria	??
Regulation of plant diversity	Arbuscular mycorrhizal fungi,	0-50% ¥
Stimulation of plant diversity	rhizobia	
Reduction of plant diversity	Arbuscular mycorrhizal fungi	20-0%§§

*Spehn *et al.*, 2002; Vogelsang *et al.*, 2006. # Note it is difficult to estimate the impact of pathogens on plant productivity in natural plant communities because other plant species increase in abundance when others decline. !! Hattenschwiler *et al.* 2005. $ DeLuca *et al.*, 2002. ¶ Hobbie and Hobbie 2006. **Houlton *et al.*, 2006. † Van der Heijden *et al.*, 2006b; the impact of P-soluble bacteria is unresolved. ¥ Van der Heijden *et al.* 2006a. §§ O_Connor *et al.*, 2002. ?? The impact of soil microbes on nitrate and phosphorus leaching is still unresolved.

grazing by protozoa and nematodes has been shown to enhance plant productivity via stimulation of nitrogen mineralization through the microbial loop. Moreover, Collembola feed on a wide range of fungi (including mycorrhizal fungi) and this can alter soil nutrient availability, plant nutrient uptake and growth, and carbon exudation into the rhizosphere, thereby altering plant-microbial competition for nitrogen.

SOIL ECOSYSTEM MANAGEMENTS

Tillage effects

The uses of deep/clean tillage techniques around tree of horticultural plant and in row spacing between plants is very common among farmers and researchers. It not only impose a physical stress on the soil structure but also on the soil microbial communities that inhabit different soil biochemical transformation process. In an effort should be made to minimize such stresses, modern arable farming systems is attempting to reduce excessive cultivation in favour of limited or strategic tillage practices.

In contrast, through conservation, or reduced tillage, can encompass any tillage practice that reduces loss of soil and water as compared to unridged or clean tillage. This can include (1) minimum tillage, considered to be the minimum amount of tillage required for plant establishment; (2) no tillage/zero tillage/direct drilling, which involves no other than chemical preparation and soil opening for plant establishment; and (3) high-residue mulched beds for plant establishment. Compared to conventional tillage systems, reduced-tillage practices offer not only long-term benefits from soil stability, reduced soil erosion especially in hilly areas, and sustainable agriculture, but they can also enhance soil microbial diversity (Davis *et al.,* 2002; Phatak *et al.,* 2002). Thus, minimizing the mechanical upheaval associated with tillage operations tends to maximize soil microbial diversity because the disruption of food substrate at the trophic level, desiccation and soil compaction are reduced, and optimum pore volume is maintained. Paradoxically, during inactive periods of main horticultural crops (*e.g.* almond and saffron crop combination), a minor crop can increase soil microbial diversity, an effect probably associated with food substrate restoration over time. Thus, heterogeneity in soil microbial populations tends to coincide with heterogeneity of food resources, which is often greatest in crops under conservation or zero tillage management, where the residue of the preceding year's crops adds sequentially to the variety of food substrates available for utilization.

Clearly, while the act of mixing soil during tillage increases homogeneity, it will simultaneously destroy the diversity of trophic microsites that occur down the soil profile together with the assemblages of soil microorganisms that occupy them. The result is a reduction in both the structural and functional diversity of the soil microbial community and the efficiency of those microbially mediated processes that sustain the horticultural productivity of soils, *e.g.*, nutrient recycling, degradation of toxic residues, maintenance of soil structure, and aggregation etc.

Bacterial endophytes and crop rotations

A continuous apoplastic pathway exits from the root epidermis to the shoot, which appears sufficient for movement of microorganisms from the root cortex into the xylem and from there throughout the plant. Consequently, many bacterial endophyte communities are the product of a colonizing process initiated by rhizo-bacteria in the root zone. In this regard, the plant host offers the microbial endophyte a relatively homeostatic, spatially diverse environment that is suitable for biotrophic mutualists, benign commensals, and necrotrophic pathogens alike.

Where plant endophyte populations are complementary, two broad categories of beneficial effect are found, namely (*i*) plant growth promotion, and (*ii*) disease control based on post infection pathogen suppression. Maintaining of plant endophyte populations is very essential, especially in those situations where endophytes are able to mitigate plant responses to environmental stress, such as those found as a result of drought, transplantation, heat etc. However, where plant endophyte populations are non-complementary, inhibitory allelopathic effects can result. It soon becomes clear that a fundamental source of potential endophytes is to be found in the soils and organic debris of previous horticultural crops. Thus, by extension, complementarity among rotation crops will involve a microbial compatibility between the newly planted crop and the resident autochthonous soil population. Consequently, it has been proposed that the yield benefits from complementary cropping systems involving, for example, legume rotations and often attributed solely to residual nitrogen and improved soil structure may also include an additional benefit from the carryover of the residual populations of endophytically competent bacteria able to promote plant growth and inhibit disease development. Accordingly, crop rotation selection criteria should include an evaluation of the effects of *rhizobacteria* and their associated endoplant competent bacterial communities when considering the long-term consequences of those crop choices, since the benefits to soil health and quality of such choices are likely to be cumulative.

Soil Ethylene and Crop Development

While the details and significance of soil ethylene (C_2H_4) have only recently become appreciated, some repercussions have already been recognized. Soil ethylene can affect plant growth and is certainly a significant factor in field production of horticultural crops. Soil is itself a source of ethylene, and the amount generated can vary widely depending on the type and degree of soil amendment and other related factors. While both biological and chemical processes contribute to ethylene accumulation in soil, most soil ethylene is produced by microbes.

Interestingly, although all major groups of bacteria, *actinomycetes,* fungi, and algae are capable of producing ethylene, spore forming bacteria living in anaerobic microsites are also primary ethylene producers. However, microbial generation of ethylene varies greatly with soil environment, the nature of substrates present in native soil organic matter, and/or soil. Smith (1976) proposed that the "microbial nutrient status" of soils could be screened by measuring O_2 utilization and CO_2 evolution and then could be correlated to soil ethylene production. In soil, the ethylene "atmosphere" may exceed 10 μL^{-1} (Perret and Koblet, 1984), sufficiently high to cause physiological changes in plants. As a gaseous plant hormone ethylene can, in very small concentrations, affect plant release from dormancy period, shoot and root growth and differentiation, adventitious root formation, leaf and fruit abscission, floral induction, female flower formation, flower leaf senescence, and fruit ripening, to mention a few of the better characterized responses.

Consistent with its classification as a secondary metabolite, microbial ethylene may also function as a signalling compound in plant microbial interactions in plant microbial-cross talk (Arshad and Frankenberger, 2002), and may also play a significant role in rhizo-bacteria mediated induced systemic resistance (ISR) in plants against specific pathogens via the jasmonate/ethylene signal-dependent pathway. Ethylene is involved in the determination of root system morphology, plant responses to abiotic and biotic stresses, and development of legume rhizobia symbiotic associations. It may also affect the rate of root infection by phytopathogens that produce ethylene during early stages of infection.

Besides soil microorganisms, plant roots contribute significant amounts of ethylene or its precursor, 1-aminocyclopropane 1-carboxylic acid (ACC), to the soil, particularly under stress. Upon exposure to stress, plants produce "stress ethylene," which can be lowered by associated bacteria able to produce 1-aminocyclopane-1 carboxylate deaminase (ACC-deaminase). The ACC-deaminase cleaves ACC to α-keto butyrate and ammonia. ACC-deaminase commonly occurs in soil bacteria, including the *pseudomonas* (Campbell and Thomson, 1996), and in rhizobia, where it stimulates the formation of root nodules in leguminous plants upon infection. Thus it is not beyond the bounds of possibility to modify crop growth by manipulation of soil ethylene level via management of ACC-deaminase containing bacterial communities.

Feeding Soil Microbes

The symbiosis that exists between plants and soil microbes is much more highly evolved and extensive than was first realized. For decades, the use of amino acid supplements, sugars, humic acids, and various gaseous materials remained outside the mainstream of agriculture in the India like country because the scientific basis of such amendments was poorly understood. However, it is increasingly apparent that many of these materials and treatments directly benefit or stimulate microbial populations by performing functions that directly benefit plants. Tailoring amendments and cultural practices to promote beneficial soil microbes has been an underappreciated area of crop production science that offers potential for increasing horticultural crop productivity in a natural and sustainable manner.

It is already well established that sugars and amino acids are released by decomposing plant material (leaf and other debris of plant) and can serve as carbon sources for soil microbes. However, in modern horticultural crop production scenario, monocultures that rely on mineral fertilizers, carbon sources can become limited, especially litter fall from plant are removed from fields and soil organic matter is kept low. Consequently, the diversity of microbial activity is likely to be reduced. This is not meant to imply that soil applications of nitrogen, phosphorus and potassium primarily intended to provide essential nutrients to crop plants do not also benefit soil microbes. The point is that traditional fertilizer inputs are intended primarily for horticultural plants and not the microbes that sustain them. Even when soil organic matter is low, relatively very few horticulturists in India would fertilize their fields specifically to benefit soil microbes. Even so, simple sugars, in the form of molasses, have often been added to compost to hasten decomposition, and on occasion they are used in small quantities by horticulturists as a soil amendment (Schenck, 2001). The benefits of sugar supplements have been reported for many years and include no detrimental environment effects because they decompose in soil to CO_2 and harmless organic products. Using sugar as a fertilizer was found to increase horticultural crop yields, particularly on soils low in potassium as prevalent in several parts of India. It has been suggested that molasses soil treatments effectively reduced crop damage by root parasites. Molasses applied to field plots at relatively high rates lowered nematode soil populations and improved papaya tree growth and harvestable fruit, and in Chinese cabbage fields it decreased the numbers of *Heterodera* nematode cysts following harvest (Schenck, 2001). Here it was reasoned that the suppressive effects of sugar on nematodes were due variously to antagonism by microorganisms in the soil that were stimulated by molasses, changes in oxygen concentration caused by microbial metabolism of molasses, or the release of toxic compounds from decomposing molasses.

Soils in clover crops have been amended with sucrose to mimic rhizosphere carbon inputs (Grayston *et al.*, 1998). After treatment, microbial communities were extracted and analysed using the BIOLOG system to construct sole carbon source utilization profiles for these communities. A clear discrimination between the carbon sources utilized by microbial communities from the different plant rhizospheres was shown (Grayston *et al.*, 1998). Accordingly, different carbon sources may stimulate specific soil microbes so that microbial fertilization can be customized to benefit particular microbes and thus the crops associated with them. Carbohydrates, carboxylic acids, and amino acids were the substrates mainly responsible for this differentiation, suggesting that plants may differ in the selective exudation of these compounds.

B. megaterium has the ability to solubilize inorganic phosphate, suppress Rhizoctonia root rot of soybean, and produce polysaccharides. In greenhouse studies, *B. megaterium* isolated from sugar treated potting soils displayed improved soil particle aggregation and increased water-holding capacity, which was considered due to his same exopolymer (Welbaum, *et al.*, 2004). Similarly, soils amended with molasses had greater moisture retention in a pawpaw field study in Australia, although the authors did not speculate on a possible mechanism (Vawdry *et al.*, 2002). Sugars may also directly impact soil properties in ways not related to microbial activity. Molasses and gypsum, applied either alone or combined can improved the structural stability sodic soils by decreasing dispersion and/or slaking. The polymer produced by *B. megaterium*, an example of microbial extracellular polysaccharides (EPS) that are apparently produced by a wide range of soil microbes (Roberson *et al.*, 1995). EPS contribute to the stability of soil aggregates and are important factors affecting soil structure in orchard of horticultural crops. In studies of tomato production in California, it was demonstrated that EPS production could be effectively managed primarily by manipulating nitrogen (Roberson *et al.*, 1995). Glomalin is another example of a beneficial microbial product produced by arbuscular mycorrhizal fungi in the taxonomic order Glomales. Glomalin is thought to seal and solidify the outside of the fungi's hyphae that transport water and nutrients, particularly phosphate, to horticultural plants. As roots grow, glomalin sloughs off into the soil, where it improves aggregation of soil particles. It can also serve as a sink for soil carbon and increases soil water holding capacity while retaining soil structure and aeration, and soil stability to resist erosion. Glomalin levels can be maintained or increased by using no till, cover crops, reduced phosphorus inputs in the field. Higher CO_2 levels in the atmosphere also stimulate arbuscular fungi to produce more glomalin. Thus at CO_2 levels of 670 $\mu L\ L^{-1}$ arbuscular mycorrhizal hyphae grew longer and produced five times as much glomalin as fungi on plants grown under ambient levels (370 $\mu L\ L^{-1}$) (Rillig *et al.*, 2000).

Apart from their trophic role, sugars also have a hormone like function as primary messengers in signal transduction in plants and certain microbes (Leo´n and Sheen, 2003). The concept of sugars as central signaling molecules is relatively novel. Hexokinase (HXK) has been found to be a glucose sensor that modulates gene expression and multiple plant hormone signaling pathways. Recent molecular analyses have revealed direct, extensive glucose control of abscisic acid biosynthesis and signaling genes that partially antagonizes ethylene signaling during seedling development under light (Leo´n and Sheen, 2003). Glucose and abscisic acid promote plant growth at low concentrations but act synergistically to inhibit growth at higher concentrations.

Sugar molecules modulate many vital processes that are also controlled by hormones during plant growth and development. For example, trehalose (α-D glucopy-ranosyl-[1,1]-α-D-glucopyranoside), a non-reducing disaccharide common in fungi and bacteria that does not accumulate in plants but is involved in the regulation of plant growth and stress responses. Transgenic plants expressing the Escherichia coli or Saccharomyces cerevisiae genes for trehalose synthesis, exhibited strong developmental alterations, such as stunted growth (Romero *et al.,* 1997). The mechanisms by which trehalose metabolism alters plant development are not entirely known. Shen and Welbaum (2004) found that 15 mM trehalose reduced both root and hypocotyl growth of alyssum and muskmelon by approximately 15%. Thus, microbially derived trehalose may similarly affect plant growth and development. The fact that sucrose, trehalose, and other sugars are not just simple molecules providing carbon and energy source but are also involved in sensing and signaling pathways both in plants and microbes adds a further degree of complexity to the functioning of the agro-ecosystem.

Techniques have been developed to map the availability of sugars and amino acids along live roots in an intact soil root matrix with native microbial soil flora and fauna present. Genetically engineered Erwinia herbicola 299R to contain a promoter less ice nucleation reporter gene, driven by either of two nutrient-responsive promoters, for use as a biosensor.

Both biosensors exhibited up to 100 fold differences in ice nucleation activity in response to varying substrate abundance in culture. Plants leak nutrients that nourish soil bacteria, which in turn benefit plants by protecting them from disease, stimulating plant growth, increasing nutrient availability, or, in the case of *B. megaterium*, secreting polysaccharides that help both the bacteria and plants avoid drought stress.

Many bio-stimulants have been offered for sale commercially in India. Although most of these products are advertised as enhancing plant growth, or increasing stress tolerance, the stimulation of bacterial growth may be an inadvertent outcome of foliar feeding. Humic substances (HS) have generally formed the major components of the mixture of materials that comprise soil organic matter. Much of the humic acid research has dealt with its creation, function, and longevity in soils. Relatively few studies have examined its effects on plant growth when used as a bio-stimulant (Kelting *et al.*, 1998). Humic acid may also be an important source of organic carbon for microbial populations, since humin is often broadly used to describe a mixture of materials that are insoluble in aqueous systems and that contain non-humic components such as long-chain hydrocarbons, esters, acids, and even relatively polar structures of microbial origin, such as polysaccharides.

Tailoring inputs to feed soil microbes has not been widely practiced in by horticulturist. As our knowledge of soil microbial communities' progresses, more precise rhizosphere management practices enhancing soil and plant health and the overall horticultural crop productivity can be urbanized.

PRECISION MANAGEMENT OF A SUSTAINABLE SOIL ECOSYSTEM: PROGRESS AND FUTURE

In general, horticultural production tends to focus on practices designed to manage suitable cultivar, training and pruning, nutrients, water, weeds, and crop diseases. With the modern integrated nutrient and pest management programs have gone one step further by recognizing the need to target inputs where they are required in the field. Here the objective has been to lower the adverse environmental impacts of intensive agriculture practices and reduce per unit production costs. However, this undoubtedly successful approach can be further refined by the development and integration of new technologies, including new crop cultivars with integration of nanotechnology that capitalize on the holistic management of the agro-ecosystem.

A better understanding of the dynamic nature of the plant-soil microbial ecosystem, the development of stable/beneficial associations between existing and newly bred crop cultivars and their associated microflora, the introduction of the new ways of microbial population management, and the development of real-time monitoring and diagnostic devices, will favor timely production interventions.

A. Plant microbial ecosystems

Certainly from the above discussion there is support for the view that the functional biodiversity of soil organisms can be linked with the maintenance of soil functional processes.

As a result, the study of soil food webs the inter-relationship between biological activity, soil micro flora, plant community structure, and soil management practices has been the focus of intense interest in recent years. Even so, our knowledge of the interaction between soil food web decomposers and their effects on nutrient cycling is generally sparse.

Perhaps even less well understood are the physiological, biochemical, plant microbial and microbial-microbial interactions that occur in real time over changing agro-physical and agrochemical environments. What are the molecular determinants of nutrient sequestration giving specific advantages to groups of different microbial consortia in various niches ? What signal molecules are involved in nutrient sensing and how are they exchanged between plants and their colonizing microflora ? How do plants recognize beneficial microorganisms and differentiate them from pathogens ? Suppression of plant defences in early stages of the rhizobia legume symbiosis development and in mycorrhizal infection indicates that such differentiation is indeed possible.

Developments in genomics, proteomics, and metabolomics provide new tools that will allow us to study microbial genotype and soil environment molecular. These tools are currently being used on model plant species, but their application to a broader range of horticultural crops is essential if we are to develop new approaches to the management of soil ecosystem and/or agro ecosystems. Breeding plants for sustainable agricultural production systems will incorporate new found disease resistance genes. However, we must also consider beneficial plant microbial associations in our breeding programs. Discovery of new small signal molecules in plant responses to environmental stresses, *e.g.,* trehalose, linear b-1, 3 glucans, fatty acid derived compounds, and microbial metabolic networks on one hand, and advances in microbial and plant genome sequencing, functional genomics and in-silico interaction of metabolic networks modeling on the other, will increase the number of breeding options and selection criteria available to plant breeders. Certainly communities of microorganisms and their host plants could benefit from closer genetic links. Interaction between soil particles and microorganisms has been recognized as one of the key determinants in a functioning soil ecosystem. The introduction of certain mycorrhizal fungi rhizo-bacteria combinations can not only affect plants but also influence rhizo-microbial composition and soil structure, enhancing, for example, the proportion of water-stable aggregates (Bethlenfalvay *et al.,* 1997).

Only a small proportion of naturally occurring microorganisms can be cultured using traditional microbiological techniques. Fortunately, molecular methods allow the monitoring of both bacteria forms and can distinguish between active and dormant constituents in any population. Moreover, a combination of molecular techniques and electron (scanning, transmission microscopy) and epifluorescence microscopy has opened up new possibilities to study bacteriophages in natural ecosystems (Asheford *et al.,* 2003). These bacterial viruses may play a significant role in the management of bacterial populations, as their numbers seem to be much higher than estimated before and may reach, on average, $1.5 \times 108\ g^{-1}$ of soil (approximately 4% of the bacterial population) (Asheford *et al.,* 2003). Such powerful analytical tools will enable us to better address agro-ecosystem management processes at the microbial level and link them to the development microbial utilization methods in plant production systems. In this regard, finding new ways of establishing stable associations between plants and beneficial organisms and understanding the molecular and biochemical mechanisms of signal recognition and transduction that occur in plant microbial interactions under different environments are most challenging study elements.

B. Soil Priming

Soil structure, its overall health, and crop yield potential can be modified by cultural practices. However, the concept of "*soil priming*" which we interpret as setting the "readiness" of a specific soil to receive a selected horticultural crops, is still new. For example, the design composition of a cover crop in a heavy life-mulch/no-till production system can include "*primer plants*" with deep, thick, and fast decomposing roots, providing channels (biopores) for the establishment of soil microbial communities. Integration of legumes, which secrete compounds recognized as signal molecules by rizo-bacteria, into life-mulch will induce the bacterial production of Nod factors (lipochito-oligosaccharides), essential for the establishment of successful N_2-fixing legume rhizobia symbiosis. Sub-micro molar concentrations of these compounds induce physiological changes in both host (legume) and an array of non-host/ horticultural plants by enhancing early seedling growth and/or plant growth. Moreover, nodules of several legumes produce high levels of hydrogen gas, which stimulates proliferation of beneficial bacteria in the rhizosphere, improving growth of cereals in rotation. Recent interest in the hydrogen producing microorganisms has been motivated by the potential for the biological production of hydrogen gas for fuel, and options also exist for their use in conventional agriculture system.

BIOPRIMING PLANT PROPAGULES

Despite decades of extensive investigation of seed inoculation with beneficial microorganisms and tuber, the full potential for the utilization of these natural allies has never been achieved. A promising new area is that of induced resistance responses in host plants mediated by selected strains of rhizosphere bacteria. These appear to protect plants by activating general internal defense responses.

This phenomenon, phenotypically similar to pathogen induced systemic acquired resistance (SAR), is referred to as rizo-bacteria mediated ISR. SAR in plants is induced by a variety of necrotizing pathogens and depends on salicylic acid (SA) accumulation in infected plants. SA-independent ISR has been well characterized in the model plant *Arabidopsis thaliana*, using *Pseudomonas fluorescence* strain WCS417r as the inducer (Ton *et al.*, 2002). ISR, is now considered a general plant defense phenomenon, and has been described in many plant species, including tobacco, tomato, cucumber, bean, radish, carnation and some horticultural crops also. The advantage of using ISR as a tool for promoting disease resistance in plants is that it provides a broad-spectrum resistance against several different types of pathogens. In *Arabidopsis*, for example, *P. fluorescence* strain WCS417r-induced ISR was effective against fungal root pathogen *Fusarium oxysporum*, and oomycete leaf pathogen

Peronospora parasitica, and bacterial leaf pathogens *Pseudomonas syringae* pv. tomato and *Xanthomonas campestris* pv. armaraciae (Pieterse *et al.*, 2000). Tissue culture techniques provide a great opportunity for the uptake of selected microbial strains and/or strain combinations by sterile (or relatively sterile) plant propagules. *In-vitro* and *ex-vitro* bacterization and mycorrhization of vegetatively propagated material is currently being explored as an efficient way to improve production practices of high value horticultural crops. The use of endophytic bacteria seems to be particularly promising (Nowak and Shulaev, 2003). The successful introduction of endophytic *Pseudomonads* into tissue culture plantlets to improve transplant establishment and early vigor has also been found to increase resistance to biotic and abiotic stresses. Similar introduction of nitrogen-fixing endophytes (other than *Rhizobia*) into clonally propagated plants could also significantly contribute to the development of sustainable crop production systems. In the tomato system, the bacterium could be transmitted through seeds; approximately 20% of seedlings derived from the seeds of Azotobacter beijerinckii-containing tomato demonstrated N_2- fixing activity two months after germination (Sz. S. Varga, personal communication). Synergistic effects of VAM fungi and diazotrophic bacteria on nutrition and growth of various crops were also seen in by scientific community and promising finding, combined with a substantial body of literature describing the presence of "*mycorrhiza helper bacteria*" in natural ecosystems needs to be explored further.

The development of a defined co-culture microbial plant micro-ecosystem to study complex plant microbial environment interactions would contribute to the overall understanding of the plant growth ecosystem and the development of new technologies for stress management and yield enhancement in plants. In this way, the large-scale production of microbial propagules, tubers, cuttings, or grafted transplants containing endophytic and/or epiphytic microbial populations designed for specific planting sites and/or crop rotations would contribute significantly to the creation of sustainable agro-ecosystems.

MICROBIAL METICULOUSNESS A SMART FIELD CONCEPT

With the continuing evolution of molecular genetic techniques, computer technology, and nanotechnology, it will be possible to precisely manage the rhizosphere while monitoring soil and plant health, in real time. Although much progress has been made in understanding those factors that comprise a sustainable agro-ecosystem and the dynamic interactions of microbial communities with soil/soil particles and plants, much more needs to be done to design "*microbial precision*" production systems.

Today, genetically based microbial profiling (Sessitsch *et al.,* 2001, 2002) allows us to link microbial population dynamics to soil type and farming inputs and outputs. Most microbial ecology studies on this topic focus on the management of rhizosphere diseases, the beneficial effects of the phylloplane organisms, *i.e., methylobacteria* and yeasts. The effects of soil and plant nutrition on resident microbial populations (both exo and endoplant) can be exploited in future scenario.

Adaptation of environmental probes based on nanosensors for agro-ecosystem monitoring would allow the development of "*smart fields*." In smart fields, simple though essential environmental factors such as temperature, pH, water, mineral nutrients, and gas composition (*e.g.*, oxygen, carbon dioxide, and ethylene) could be monitored in real time and at multiple locations in the field. Such probes could be integrated into arrays and linked to a computer based "monitoring/command" system that turns on specifically prescribed response inputs to correct for imbalances that might arise. Changes to the rhizosphere could be rapidly made through a subsurface irrigation system buried to a depth of 15-25 cm. Diverse inputs, such as water, mineral nutrients, microbial inoculants, crop protection chemicals, gases (*e.g.*, O_2, air, or H_2) and organic nutrients (*e.g.,* sugars and/or signal molecules) that preferentially feed soil and plant beneficial microbes could readily be delivered, creating both beneficial rhizospheres and healthy plant stands. Much more attention could be paid to field site preparation prior to planting (soil priming) to reduce soil-borne disease pressures and thus post planting inputs. Smart Field technology, combined with microbial population analyses dynamics, offers the possibility of (1) further improving our understanding of interactions within the soil microbial–plant ecosystem; (2) better enabling the identification of indicators of soil and plant health status, allowing for early interventions and thus reducing the requirement for inundative (release of overwhelming numbers of a mass-produced) system inputs; (3) creating a scientific "workshop" for the development of new sustainable technologies for accelerated soil improvement; and (4) provide impetus for discovery and the development of simple diagnostic tools for the determination of soil and plant health status parameters.

REFERENCES

Alexander, I.J., Lee, S.S. (2005). Mycorrhizas and ecosystem processes in tropical rain forest : implications for diversity. In : Biotic Interactions in the Tropics : Their Role in the Maintenance of Species Diversity (eds Burslem, D.F.R.P., Pinard, M.A. and Hartley, S.E.). Cambridge University Press, Cambridge, pp. 165-203.

Arshad, M. and Frankenberger, W.T., Jr. (2002). Ethylene, Agricultural Sources and Applications. Kluwer Acad. Publ. /Plenum Publ., New York.

Ascheford, K.E., Day, M.J., and Fry, J. 2003. Elevated abundance of bacteriophage infecting bacteria in soil. Appl. Envir. Microbiol. **69** : 285-289.

Bardgett, R.D., Bowman, W.D., Kaufmann, R., Schmidt, S.K. (2005). Linking aboveground and belowground communities: a temporal approach. Trends Ecol. Evol., **20** : 634-641.

Bethlenfalvay, G.J., Andrade, G., and Azc´on-Aquilar, C. 1997. Plant and soil responses to mycorrhizal fungi and rhizobacteria in nodulated or nitratefertilized peas (Pisum sativum L.). Biol. Fertil. Soils. **24** : 164-168.

Blagodatsky, S.A., Richter, O. Microbial growth in soil and nitrogen turnover: a theoretical model considering the activity state of microorganisms. Soil Biol Biochem.1998, **30** : 1743-1755.

Bonner, J. and J.E. Varner. 1967. Plant Biochemistry, p. 199. Academic Press, New York, NY.

Brooks, P.D., Williams, M.W., Schmidt, S.K. (1998). Inorganic nitrogen and microbial biomass dynamics before and during spring snowmelt. Biogeochemistry, **43** : 1-15.

Campbell, R.B. and Thomson, J.A. 1996. 1-Aminocyclopropane-1carboxylate deaminase genes from Pseudomonas strains. FEMS Microbiol. Lett. **138** : 207-210.

Chesbro, W. The domains of slow bacterial growth. *Canadian Journal of Microbiology.* 1988, **34** : 427-435.

Cleveland, C.C., Townsend, A.R., Schimel, D.S., Fisher, H., Howarth, R.W., Hedin, L.O. *et al.* (1999). Global patterns of terrestrial biological nitrogen (N-2) fixation in natural ecosystems. Global Biogeochem. Cycles, **13** : 623-645.

Davis, L.C., Castro-Diaz, S., Zhang, O., and Erickson, L.E. (2002). Benefits of vegetation for soils with organic contaminants. Crit. Rev. Plant Sci. **21** : 457-491.

DeLuca, T.H., Zackrisson, O., Nilsson, M.C. and Sellstedt, A. (2002). Quantifying nitrogen fixation in feather moss carpets of boreal forest. Nature, **419** : 917-920.

Gans, J., Wolinsky, M. and Dunbar, J. (2005). Computational improvements reveal great bacterial diversity and high metal toxicity in soil. Science, **309** : 1387-1390.

Grayston, S.J., Wang, S., Campbell, C.D. and Edwards, A.C. 1998. Selective influence of plant species on microbial diversity in the rhizosphere. Soil Biol. Biochem. **30** : 369-378.

Hattenschwiler, S., Tiunov, A.V. and Scheu, S. (2005). Biodiversity and litter decomposition in terrestrial ecosystems. Ann. Rev. Ecol. Evol. Syst., **36** : 191-218.

Heijden M.G. Bardgett R.D. and Straalen N.M. (2008). The unseen majority: soil microbes as drivers of plant diversity and productivity in terrestrial ecosystems. Ecology Letters, **11** : 296-310.

Helgason, T., Daniell, T.J., Husband, R., Fitter, A.H. and Young, J.P.W. (1998). Ploughing up the wood-wide web? Nature, **394** : 431-431.

Hobbie, J.E. and Hobbie, E.A. (2006). N-15 in symbiotic fungi and plants estimates nitrogen and carbon ?ux rates in Arctic tundra. Ecology, **87** : 816-822.

Hodge, A., Campbell, C.D. & Fitter, A.H. (2001). An arbuscular mycorrhizal fungus accelerates decomposition and acquires nitrogen directly from organic material. Nature, **413** : 297-299.

Hodges, C.F. and Campbell, D.A. 1998. Gaseous hydrocarbons associated with black layer induced by the interaction of cyanobacteria and Desulfovibrio desulfuricans. Plant Soil **205** : 77-83.

Houlton, B.Z., Sigman, D.M. and Hedin, L.O. (2006). Isotopic evidence for large gaseous nitrogen losses from tropical rainforests. Proc. Natl Acad. Sci. USA, **103** : 8745-8750.

Jetten, M.S.M., Stams, A.J.M., Zehnder, A.J.B. Acetate threshold values and acetate activating enzymes in methanogenic bacteria. FEMS Microbiol Ecol, 1990. **73** : 339-344.

Jonsson, L.M., Nilsson, M.C., Wardle, D.A. and Zackrisson, O. (2001). Context dependent effects of ectomycorrhizal species richness on tree seedling productivity. Oikos, **93** : 353-364.

Kjelleberg, S., Hermansson, M., Marden, P., Jones, G.W. The transient phase between growth and nongrowth of heterotrophic bacteria, with emphasis on the marine environment. Annu Rev Microbiol 1987, **41** : 25-49.

Kucey, R.M.N. (1983). Phosphate solubilzing bacteria and fungi in various cultivated and virgin Alberta soils. Can. J. Soil Sci., **63** : 671-678.

Le´on, P. and Sheen, J. (2003). Sugar and hormone connections. Trends Plant Sci. **8** : 110-116.

Lozupone, C.A. and Knight, R. (2007). Global patterns in bacterial diversity. Proc. Natl Acad. Sci. USA, **104** : 11436-11440.

Marr, A.G., Nilson, E.H., Clark, D.J. The maintenance requirements of *Escherichia coli*. Ann N Y Acad Sci 1963, **102** : 536 -548.

Monod, J. The growth of bacterial cultures. *Annual Review Microbiology.* 1949, **3**: 371-39.

Nowak, J. and Shulaev,V. 2003. Priming for transplant stress resistance in in vitro propagation. In Vitro Cell. Dev. Biol.-Plant **39** : 107-124.

O' Connor, P.J., Smith, S.E., Smith, F.A. Arbuscular mycorrhizas influence plant diversity and community structure in a semiarid herb land. *New Phytol.*, 2002, **154** : 209-218.

Panikov, N.S. Microbial Growth Kinetics, 1st edn. 1995. Chapman and Hall, London.

Perret, P. and Koblet, W. (1984). Soil compaction induced iron-chlorosis in grape vineyards: presumed involvement of exogenous soil ethylene. J. Plant Nutrit. **7** : 533–539.

Phatak, S.C., Dozier, J.R., Bateman, A.G., Brunson, K.E., and Martini, N.L. (2002). Cover crops and conservation tillage in sustainable vegetable production. In: Making Conservation Tillage Conventional: Building a Future on 25 Years Research, pp. 401-403. van Santen, E., Ed., Proc. 25th Annual Southern Conservation Tillage Conference for Sustainable Agriculture, Auburn, AL. 24-26, June 2002.

Pieterse, C.M.J., Van Pelt, J.A., Ton, J., Parchmann, S., Mueller, M.J., Buchala, A.J., Metraux, J.P., and Van Loon, L.C. (2000). Rhizobacteria-mediated induced systemic resistance (ISR) in Arabidopsis requires sensitivity to jasmonate and ethylene but is not accompanied by an increase in their production. Physiol. Mol. Plant Pathol. **57** : 123-134.

Pirt, S.J. Maintenance energy: a general model for energy limited and energy sufficient growth. *Achieves of Microbiology*. 1982, **133** : 300-302.

Powell, E.O. The growth rate of microorganisms as a function of substrate concentration. In : Powell, EO, Evans, CGT, Strange, RE, Tempest, DW (Eds.) Microbial Physiology and Continuous Culture 1967. HMSO, London, pp 34-56.

Rai, A.N., Soderback, E., Bergman, B. (2000). Cyanobacterium plant symbioses. New Phytol., **147** : 449-481. Smith, S.E. & Read, D.J. (1997). Mycorrhizal Symbiosis, 2nd edn.Academic Press, London, UK.

Rillig, M.C., Hernandez, G.Y., and Newton, P.C.D. (2000). Arbuscular mycorrhizae respond to elevated atmospheric CO_2 after long-term exposure: evidence from a CO_2 spring in New Zealand supports the resource balance model. Ecology Lett. **3** : 475-478.

Roberson, E.B., Sarig, S., Shennan, C., and Firestone, M.K. (1995). Nutritional management of microbial polysaccharide production and aggregation in an agricultural soil. Soil Sci. Soc. Am. J. **59** : 1587-1594.

Romero, C., Belles, J.M., Vaya, J.L., Serrano, R., and Culianez-Macia, F.A. (1997). Expression of the yeast trehalose-6-phosphate synthase gene in transgenic tobacco plants: pleiotropic phenotypes include drought tolerance. Planta **201** : 293-297.

Russell, J.B., Cook, G.M. Energetics of bacterial growth: balance of anabolic and catabolic reactions. Microbiol Rev 1995, **59** : 48- 62.

Schenck, S. 2001. Molasses soil amendment for crop improvement and nematode management. Hawai Agricultural Research Center Vegetable Report 3, HARC, Aiea, HI.

Schlesinger, W.H. (1997). Biogeochemistry: An Analysis of Global Change, 2nd edn. Academic Press, San Diego. Tiedje, J.M. (1988). Ecology of denitrification and dissimilatory nitrate reduction to ammonium. In: Biology of Anaerobic Microorganisms (ed. Sehnder, A.J.B.). Wiley, New York, pp. 179-244.

Schmidt, I.K., Michelsen, A., Jonasson, S. (1997). Effects of labile soil carbon on nutrient partitioning between an arctic graminoid and microbes. Oecologia, **112** : 557-565.

Servais, P, Billen, G, Vives Rego, J (1985). Rate of bacterial mortality in aquatic environments. Appl Environ Microbiol. **49** : 1448-1454

Sessitsch, A., Rieter, B., Pfiefer, U., and Wilhelm, E. 2002. Cultivationindependent population analysis of bacterial endophytes in three potato varieties based on eubacterial and Actinomycetes-specific PCR of 16S rRNA genes. FEMS Microbiol. Ecol. **39** : 23-32.

Sessitsch, A., Weilharter, A., Gerzabek, M.H., Kirchmann, H., and Kandeler, E. (2001). Microbial population structures in soil particle size fractions of a long-term fertilizer ?eld experiment. Appl. Envir. Microbiol. **67** : 4215-4224.

Shen, Z.X. and Welbaum, G.E. (2004). Exogenous trehalose inhibits hypocotyls elongation of alyssum and muskmelon seedlings. Acta Hort. **63** : 135-139.

Smith, S.E. and Read, D.J. (1997). Mycorrhizal Symbiosis, 2nd edn. Academic Press, London, UK.1

Spehn, E.M., Scherer-Lorenzen, M., Schmid, B., Hector, A., Caldeira, M.C., Dimitrakopoulos, P.G. *et al.* (2002). The role of legumes as a component of biodiversity in a cross-European study of grassland biomass nitrogen. Oikos, **98** : 205-218.

Sprent, J.I. (2001). Nodulation in Legumes. Royal Bot. Gardens, Kew, UK.

Stouthamer, A.H., Bulthuis, B.A., van Verseveld, H.W. Energetics of growth at low growth rates and its relevance for the maintenance concept. In: Poole, RK, Bazin, MJ, Keevil, CW (Eds.) 1990. Microbial Growth Dynamics. IRL Press, Oxford, Special Publications Society for General Microbiology, vol. 28, pp 85-102.

Tempest, D.W., Neijssel, O.M. (1984). The status of YATP and maintenance energy as biologically interpretable phenomena. Annu Rev Microbiol. 38 : 459-486.

Tiedje, J.M. (1988). Ecology of denitrification and dissimilatory nitrate reduction to ammonium. In: Biology of Anaerobic Microorganisms (ed. Sehnder, A.J.B.). Wiley, New York, pp. 179-244.

Ton, J., Van Pelt, J.A., Van Loon, L.C., and Pieterse, C.M.J. (2002). Differential effectiveness of salicilate-dependent and jasmonate/ethylene-dependent induced resistance in Arabidopsis. Mol. Plant-Microbial Interact. 15 : 27-34.

Van der Heijden, M.G.A., Bakker, R., Verwaal, J., Scheublin, T.R.,Rutten, M., van Logtestijn, R. *et al.* (2006a). Symbiotic bacteria as a determinant of plant community structure and plant productivity in dune grassland. FEMS Microb. Ecol., **56** : 178-187.

Van der Heijden, M.G.A., Streitwolf-Engel, R., Riedl, R., Siegrist, S., Neudecker, A., Ineichen, K. *et al.* (2006b). The mycorrhizal contribution to plant productivity, plant nutrition and soil structure in experimental grassland. New Phytol., **172** : 739-752.

Varga, Sz. S., Kor´anyi, P., Preininger, E., and Gyurj´an, I. 1994. **Artificial associations between Daucus and nitrogen-fixing Azotobacter cells in vitro. Physiol. Plant. 90** : 786-790.

Vogelsang, K.M., Reynolds, H.L. and Bever, J.D. (2006). Mycorrhizal fungal identity and richness determine the diversity and productivity of a tallgrass prairie system. New Phytol., **172** : 554-562.

Welbaum, E.G., Sturz , Antony, V.S., Dong, Z., and Nowak, J. (2004). Managing Soil Microorganisms to Improve Productivity of Agro-Ecosystems, Critical Reviews in Plant Sciences, **23** : 2, 175-193.

●●●